MW00978418

The Deep Green Planet

The Breathing Earth

The Deep Green Planet

THE BREATHING EARTH

RENATO MASSA
WITH
MONICA CARABELLA AND LORENZO FORNASARI

Austin, Texas

Published by Raintree Steck-Vaughn Publishers, an imprint of Steck-Vaughn Company

Editors
Caterina Longanesi, Linda Zierdt-Warshaw
Design and layout
Jaca Book Design Office

Library of Congress Cataloging-in-Publication Data

Massa, Renato.
[Terra che respira. English]
The breathing earth / Renato Massa with Monica Carabella and Lorenzo Fornasari.
p. cm. — (The Deep green planet)
Includes index.
Summary: Provides a basic introduction to the science of ecology by discussing natural recycling, photosynthesis, and the chemistry of life.
ISBN 0-8172-4309-7
1. Ecology — Juvenile literature. 2. Photosynthesis — Juvenile literature. [1. Ecology. 2. Photosynthesis.] I. Carabella, Monica. II. Fornasari, Lorenzo, 1960– . III. Title. IV. Series.
QH541.14.M3613 1997
574.5 — dc20 96–12836
CIP AC

Printed and bound in the United States
1 2 3 4 5 6 7 8 9 0 WO 99 98 97 96

Contents

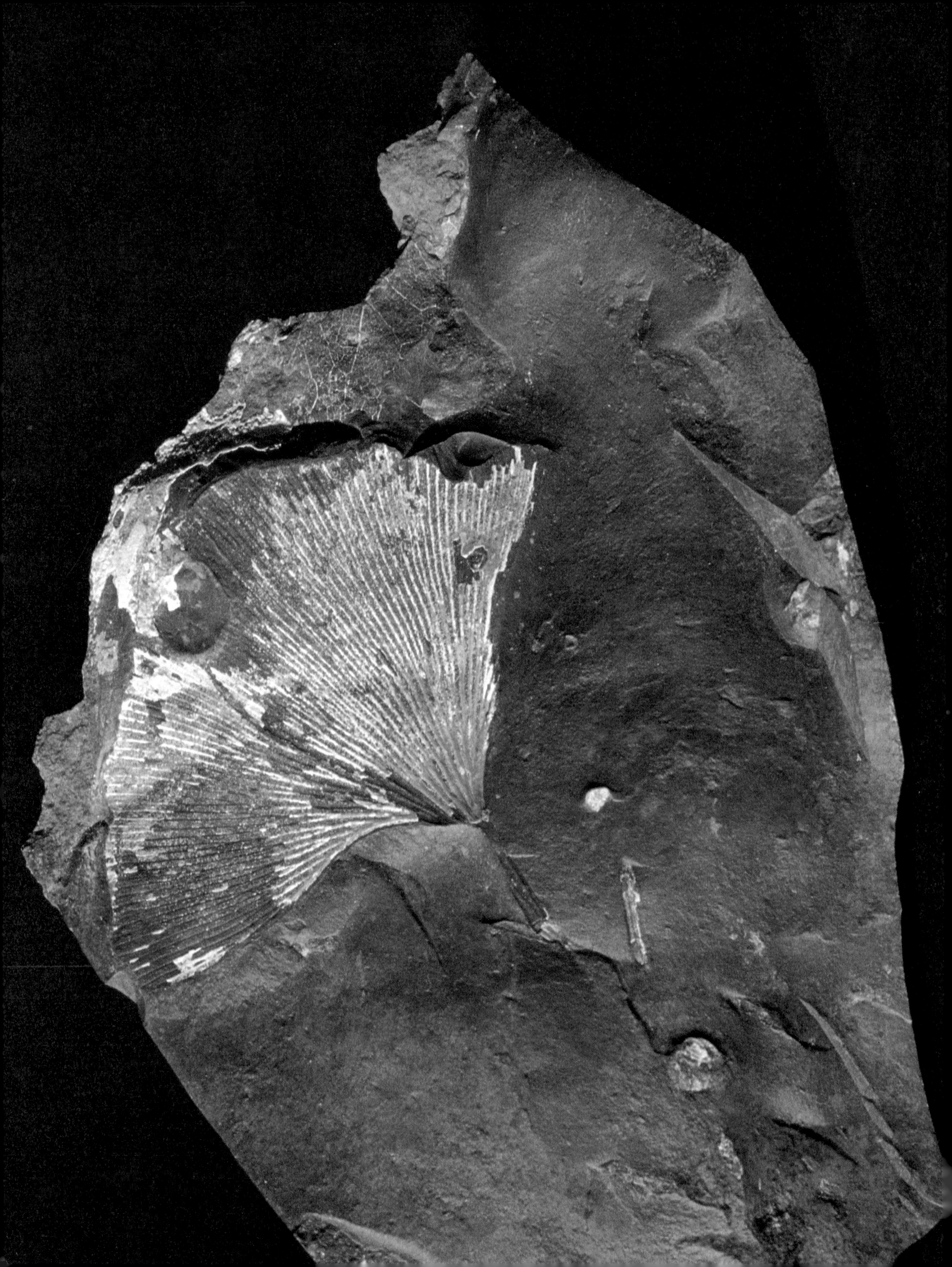

INTRODUCTION

We know of only one planet that lives and breathes. This planet is Earth. As far as we know, it is the only planet of its kind in the whole universe. However, there may be some similar planets waiting to be discovered.

About four billion years ago, molecules that could duplicate themselves through a simple chemical reaction appeared. The duplicated molecules developed to a point where they no longer resembled the original molecules. Over time, these molecules developed methods of protecting themselves from harmful chemicals. Pieces of these destroyed chemicals were absorbed by the original molecules and used to reproduce. At the same time, they built up defenses against harmful molecules. The result of this extraordinary activity of attack and defense was living organisms. Today, billions of organisms belonging to millions of different species live on Earth.

All organisms need energy for their activities. The sun is the main source of most of this energy. Through the process of photosynthesis, some organisms use sunlight to meet their energy needs. These organisms use the energy to construct new living material. This material and energy is taken from them by other organisms, called heterotrophs. Heterotrophic organisms are those that get their energy by eating other organisms.

The organisms that photosynthesize use the sugars they make. Heterotrophic organisms produce energy in the form of ATP molecules, starting with sugars taken from these organisms. In this way, the Earth breathes. The scientific term for this process is respiration, which is explained in the chapter entitled "How the Earth Breathes." The grasslands, the forests, and the savannas breathe. The oceans breathe. Wild and domestic animals and human beings also breathe.

The Earth has breathed as long as it has supported life. It has breathed in an ocean of bacteria. It has breathed in a world of dinosaurs and ammonites. It has breathed in a world of bison, bears, and lions. It will breathe tomorrow in a world increasingly covered with asphalt. The forests may disappear, the butterflies, birds, and squirrels may become distant memories, but as long as Earth continues to breathe, there is hope. The whole cycle of life could start over from a handful of seeds. The destiny of the Earth lies in its ability to breathe.

RENATO MASSA

What Is Ecology?

Ecology is often defined as the science that studies the relationships between living things and their surroundings. Today, scientists tend to present information in ecology as numbers. The numbers use ratios to show relationships between environmental characteristics and the **population density** of a species. Population density is the number of living things in a given area. In 1978 Nobel Prize winning biochemist Hans A. Krebs redefined the term *ecology* to address this trend. According to the definition developed by Krebs, ecology is the scientific study of the interactions that determine the distribution and population sizes of living things.

c

d

e

1

2

The two-page diagram shows all the factors that, according to Krebs' definition of ecology, decide the distribution and population size of a species in an environment.
1. The first column shows several living things.
2. The second column shows environments in which the living things pictured can survive.
3. Factors that prevent each organism pictured from living in the environments shown are listed on page 11.
a. The American robin (*Tordus migratorius*) cannot cross the Atlantic Ocean to reach Great Britain. **b.** Monk seals (*Monachus monachus*) do not live on the Mediterranean coast. The seals are not absent due to problems of dispersal. They are absent because they cannot tolerate disturbances caused by boat engines. **c.** The mountain sparrow (*Passer montanus*) can reach European city parks. However, the species cannot survive there because it cannot compete with the common house sparrow (*Passer domesticus*) for resources. **d.** This elm species, *Ulmus glabra*, no longer lives in the forests of Central Europe. Its absence is due to disease. **e.** The myrtle (*Myrtus communis*) is a Mediterranean plant. It cannot survive at high mountain altitudes because of the climate. **f.** The brown trout (*Salmo trutta fario*) does not live in river estuaries. It cannot tolerate the high level of dissolved salt and low oxygen levels of estuary waters.

Dispersal	Behavior	Interaction with Other Species	Parasitism/Disease	Physical Factors	Chemical Factors
The area is not accessible.	The habitat is not suited to the species' lifestyle.	A similar species lives in the area.	The organism's resistance to disease is inadequate.	The climate is unsuitable.	Too little or too much of some chemicals are present.

3

The green lizard (*Lacerta viridis*) shown in figure **1** and photo **3** is one of Europe's largest lizards. Its length may exceed 40 centimeters (16 inches). The species lives in the Bourne gorges of the Vercours region of southeastern France, shown in photo **2.** In this region the lizard is not exposed to any of the factors that often limit the distribution of living things.

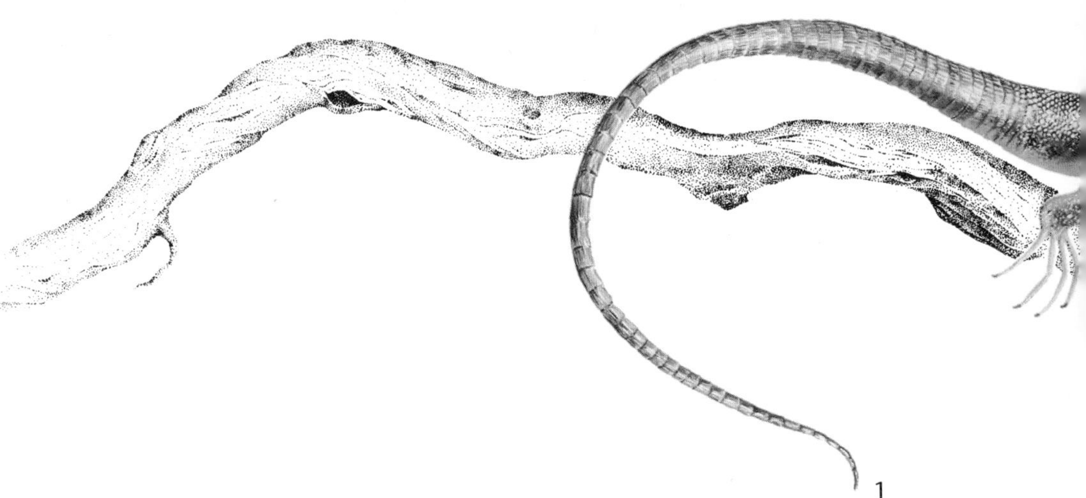

1

The Link with the Environment

Many environmental factors affect the ability of a species to adapt to a **habitat.** Plants are most affected by physical and chemical environmental factors. For example, light, temperature, and moisture clearly play a role in plant growth.

Some plants are suited to extreme environmental conditions. **Succulents,** for example, are plants with thick stems that store water. These plants are suited to the dryness and large temperature ranges of many deserts. The **tundra** is a cold, dry environment located in the far north. Tundra grasses are adapted to the short growing season of this cold, dry region. In contrast, the plants that make up the **undergrowth** of **rain forests** are suited to conditions of constant shade and dampness.

Soil is as important to plants as climate. Plant **germination** and growth depend on the chemicals in soil. Soil texture and structure are also important. These factors affect the ability of plants to root. In turn, their ability to root affects the absorption of water and mineral salts.

For each environmental factor, each species has a **range of tolerance** within which it can survive. A species also has a point at which it reaches its maximum population density. At this point, the available resources cannot provide for any more **organisms.** According to the law of the minimum, the environmental factor that is least satisfactory to a species' needs limits the presence or growth of the species. Any environmental factor that restricts the presence or growth of a species is a **limiting factor.**

Agents of Change

Plants do not play a passive role in the environment. As their roots penetrate or push through soil, plants change soil structure. Plants also change the chemical makeup of soil. They do this by adding **organic** substances to the soil and removing nitrogen compounds. Changes in soil features and the plants themselves may allow other organisms to live in the area.

Heavy plant cover can change the moisture and temperature levels in an area. Such changes create **microclimates.** A microclimate may have more humidity and a lower temperature than the bare ground and surrounding air. These effects, and interactions between organisms, such as predation, competition, and **parasitism,** are **biotic,** or living, **factors.**

2

DISPERSAL	BEHAVIOR	INTERACTION WITH OTHER SPECIES	PARASITISM/DISEASE
The area is accessible.	The habitat is suitable for the species' lifestyle.	Other very similar species do not already occupy the environment.	The organism is adequately resistant.
The Vercours area is accessible to the green lizard. It is found in steady numbers throughout south-central Europe. Thus, the lizard occupies all the areas in which its ecological needs can be met.	The broadleaf woodland is suited to the green lizard. Dense vegetation provides a place for nesting and a fairly humid environment. The forest also has open spaces in which the lizard can bask in the sun and hunt for the insects on which it feeds.	No other reptiles that occupy the same ecological niche as the green lizard and compete for the same resources live in the Vercours valleys. In addition, no predators so specialized and numerous in the Vercours valley represent a threat to the lizard.	There are no parasites or disease-causing agents in the Vercours region tha are severe enough to increase the death rate of the green lizard to a point that threatens the survival of the species.

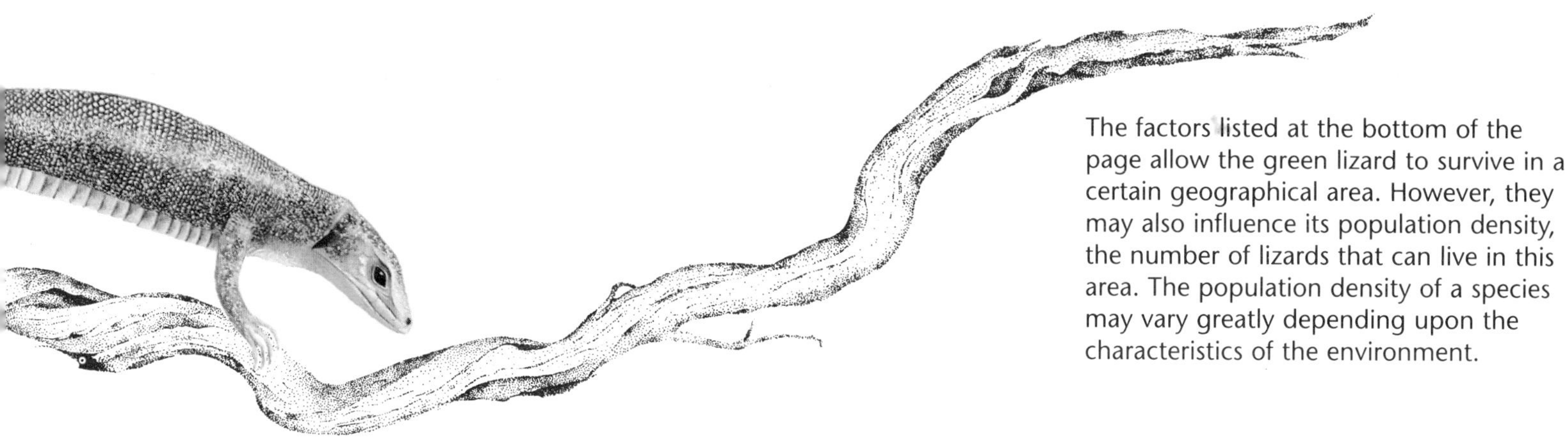

The factors listed at the bottom of the page allow the green lizard to survive in a certain geographical area. However, they may also influence its population density, the number of lizards that can live in this area. The population density of a species may vary greatly depending upon the characteristics of the environment.

Predation and Competition

In ecology, predation includes animals eating other animals and animals eating plants. Plants and animals respond differently to predation. For example, animals may avoid predation through escape or **camouflage**. Plants may use chemicals as a **defense mechanism**. Such responses may involve unpleasant tastes or poisonous substances. Other plant defense mechanisms may include spines, such as those of a cactus, or shells, such as the covering of coconuts. However, most plants avoid predation through **regeneration**. Regeneration is the regrowth of lost parts. Regeneration may occur even when a plant is severely harmed.

The presence of prey and **predators** affects the distribution of plant and animal species. The competition between species also limits distribution. Animals often compete for similar resources, such as food. They may also compete for sites suitable for reproduction or the raising of young. Plants may compete for living space, sunlight, water, **nutrients**, and pollinating insects.

In a given area, two species cannot use the same resources in exactly the same way. When such competition occurs, the less efficient species must either change the way it uses the resource or leave the area. If the species does not do one of these things, it will die. Often, competing species "split" the use of a resource. An example of how two species split the use of a resource is shown by the marsh warbler (*Acrocephalus palustris*) and the great reed warbler (*A. arundinaceus*). These species are closely related. Both species nest in reed beds. However, the marsh warbler nests in the center of the bed, while the great reed warbler nests at the edge of the bed.

Ecological Equivalents

Geographical barriers, such as an ocean or a high mountain range, may affect species distribution. Such barriers can result in similar species living in separate areas. Though separate, the species will use the same resources in the same way. Because they use resources in the same way, these species are called **ecological equivalents**.

Physical Factors
The climate is suitable.

Chemical Factors
Neither too little nor too much of some chemicals are present.

The Vercours climate, even with seasonal changes and low winter temperatures not favorable to reptiles, is suitable for the daily life of a reptile such as the green lizard. Climate changes are aptly timed for the hatching of its eggs, at least at altitudes up to 1,300 to 1,400 meters (4,265 to 4,600 feet).

The green lizard is not especially sensitive to "fine" chemical factors, such as the makeup of rocks and the acidity of soil. The species is also not so specialized that it cannot change food sources if chemical factors affect the population density of one of its prey species.

3

Ecosystems

A representation of the parts of an oasis in the Sahara, a typical ecosystem. The first level of ecological complexity is a population — all the members of a single species. Populations mix and interact in an area to form communities. The communities and the physical environment that supports them form different land ecosystems. Each ecosystem involves the interaction of living and nonliving things. This process is sustained by the supply of energy.

Populations and Communities

Ecologists study living things at different levels. The lowest level is the population. A population is all the members of the same species living in an area at the same time. The next level is the **community**. A community is all the populations that live in the same area. The community with its surroundings make up the third level—the **ecosystem**. The ecosystem includes all the living and nonliving parts of the environment and the relationships between these parts.

Each level in an ecosystem has its own traits. These traits help us understand how the ecosystem works. During its life, a living thing is born, uses energy, reproduces, and dies. An organism is the basic unit of the population. A population shares many traits with its individual members. For example, like individual organisms, populations have birthrates and death rates. Populations also change through immigration and emigration.

A community is characterized by the relationships between its populations. Predation and competition are two such relationships. These relationships direct the flow of energy between the links of the **food chains** in a community.

Evolution Toward a Climax Community

Ecosystems evolve toward stable forms. Changes in ecosystems from less stable to more stable forms are called **succession**. In ecosystems, stability is decided mostly by climate. For example, in areas where temperature and humidity are not limiting factors, forest ecosystems develop. As one moves from the poles toward the equator, these forest ecosystems become more complex. Forest ecosystems represent the final stage, or **climax**, of succession.

Successions begin with bare soil. In time, a community of pioneer plant life, such as mosses and lichens, develops. As soil becomes richer, it may support a **grassland**. The grassland may be replaced by **scrub**. Next, wooded scrub and finally a forest may develop in the area.

Each succession has different types and layers of plant life. As the types and layers of plant life increase, the communities become more complex. Living things, both plant and animal, tend to increase in size and in number. At the same time, competition between species also increases.

In a developing ecosystem there is little competition. As a result, the plant life in these ecosystems thrives, and they reproduce in great numbers. Often, large amounts of organic matter form in the area. In contrast, competition in mature ecosystems is very great. As a result, population sizes stay at about the same numbers. The energy gained through **photosynthesis** is used to maintain the ecosystem.

Climax communities decided by the physical and chemical traits of soil are **edaphic climaxes**. In different areas, the beginning stages of succession may vary based on the features of the soil. For example, acidic rocks, basic rocks, and waterlogged soils support different types of plants. The plants, in turn, decide what other living things can live in the community. Surprisingly, differing soils often produce the same final stage of succession — a **territorial climax**.

Mountains or soils of specific types may support plant species other than those you would expect to find. Forest types differ as one moves from high to low altitudes. High altitudes have the narrow-leaved forests of the cold temperate zones, called the **taiga**. Slightly below this altitude are the broadleaf **deciduous forests**. Next are the dry climate forests, the warm tropical forests, and the equatorial rain forests. These broad groups of plant life are called **formations**. Within each formation, different plant populations develop. For example, the taiga may be woodland made up mostly of firs, larches, pines, or birches.

Plant Associations

The study of the makeup of the plant life in an area has led to the idea of **associations**. These are plant groups with fairly constant compositions. In associations certain **dominant species** are linked with other characteristic species. The characteristic species are often rare. However, the species is unique or almost unique to the association.

A complex classification system of associations exists. This classification system groups associations with similar makeups and structures into higher categories. For example, one group in this class is the broad oak-beech (*Querco-Fagetea*) class. This class contains durmast oak (*Quercus pubescens*), hornbeam (*Carpinus betulus*), and manna ash (*Fraxinus ornus*) of the woodlands of hills and mountains and mixed woodlands dominated by beeches (*Fagus sylvatica*) and white poplars (*Populus alba*). Each of these plant groups is representative of different associations.

Where the plants live in an area is as important to the associations as the species makeup. All forests have at least three layers:

Tropical rain forest

Seasonal tropical forest

Savanna, with thorny bushes

Desert

herb, shrub, and tree. In complex environments additional layers may exist. For example, a moss and a **liana** layer may also be present. This layer contains the **epiphytes** that grow on trees. The amount of associability of these species can be determined layer by layer or in total. The amount of space these plants occupy can also be determined layer by layer or in total. Degree of associability is calculated considering both the foliage coverage and the numbers of the species.

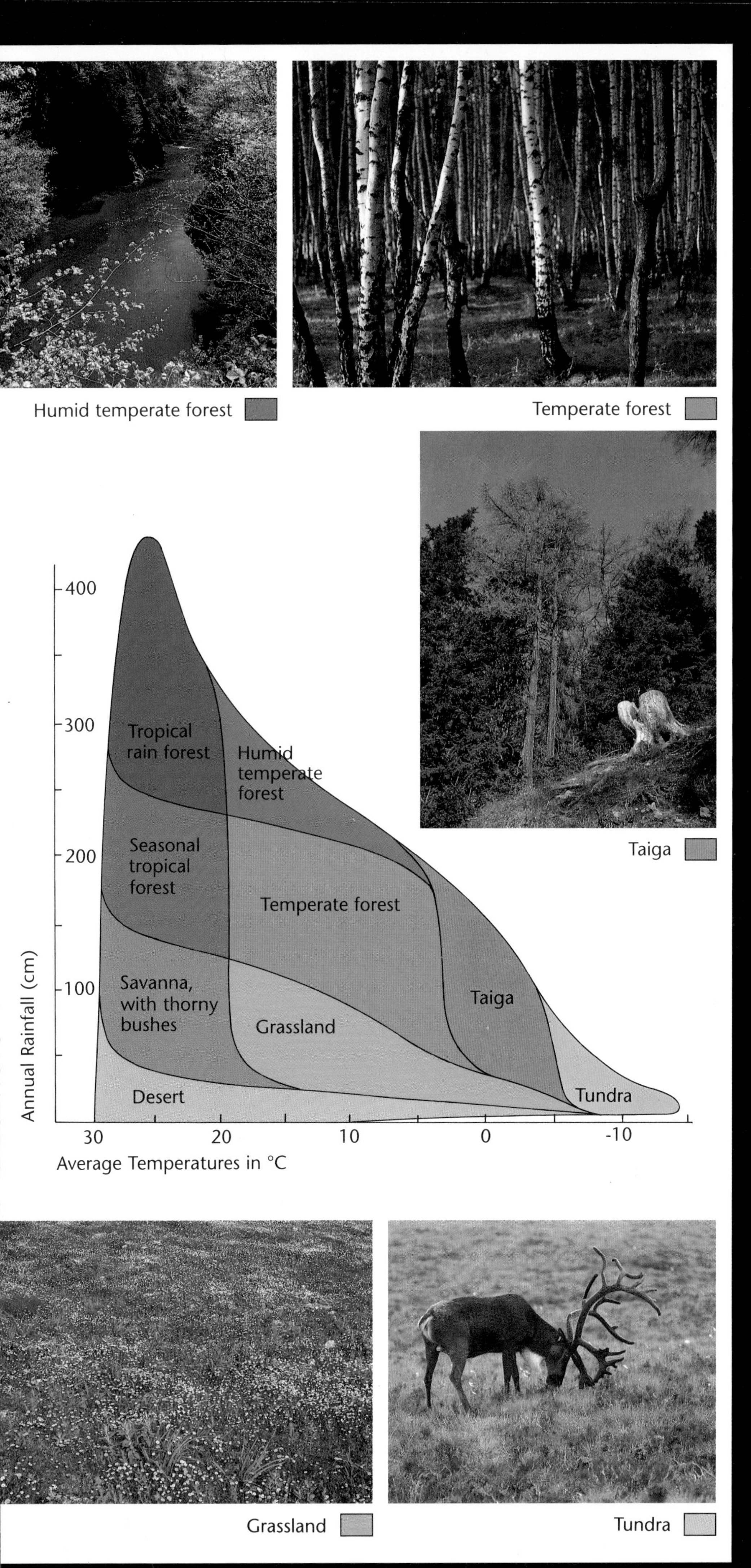

The types of plant life resulting from the range of average annual temperatures. The x axis shows degrees Celsius on a decreasing scale. Rainfall levels are shown in centimeters on an increasing scale on the y axis. As both temperature and rainfall increase, ecosystems change from desert and prairie to various types of forests. The highest temperatures and rainfall produce tropical rain forests. Apart from the climate factors, the soil also affects the type of plant life in an area.

MEASURING POPULATIONS

The Importance of Censuses

Numbers are often useful when trying to understand the relationships between living things and their environments. The numbers of individuals in an ecosystem change as environmental factors change. These changes are often compared by looking at different areas over the same period of time. The results of these studies provide data about the population densities of the areas.

The most direct way to find population density is to carry out a **census**. In a census, well-defined methods are used to count organisms. Data from different censuses is then compared. These comparisons show changes in population sizes and makeup.

The aim of a census is to get as correct a population count as possible. However, it is not always possible to count each individual in an area. Exact counts are easiest in small areas. They are also easier when counting large plants and animals. Accurate counts become difficult in large areas or as living things become smaller. For example, the number of trees in an acre of forest can be counted. However, it is almost impossible to count the number of acorns that have fallen to the ground and germinated in that acre. It is also difficult to count the total number of trees in the whole forest. In the same way, it is fairly easy to count the number of chamois in an alpine region. However, counting the number of grasshoppers in the meadows of a whole mountain would be impossible.

Sampling and Relative Densities

Scientists have two shortcuts for counting populations. The first is **sampling**. The second is **relative densities**.

In sampling, census counts are made in small areas. These counts provide density numbers for that area. The numbers are then multiplied by the size of the whole area under study to provide an estimate of the population density of the whole area.

This method is shown in the following example. Scientists want to find the deer population in a 1,000-acre forest. A count of the deer on one acre is made. If ten deer are counted, the deer population for the whole forest is estimated to be 10,000 deer. This calculation is made by multiplying the number of deer in one acre of forest (ten) by the total size of the forest (1,000 acres).

The relative density method is used to show how a population changes over time or in different environments. It does not provide an exact population size. Relative density censuses use partial counts made during certain periods of time or along migration routes. The population numbers are not tied to an entire area. Relative density figures are reported based on the time and/or the energy spent carrying out the survey. For example, they may show the number of crows seen in one hour from a specific place. You might count the number of antelope seen during a ten-mile drive across the **savanna**.

Relative censuses are often used to count birds. Such censuses allow the measurement of the change in population density from one season to another. A comparison between the densities of nesting pairs in different environments can also be made.

Indirect Measurements of Density

Censuses of animals are often based on indirect evidence. For example, mammals are often counted based on the discovery of their tracks, **feces**, or lairs. The amount of this evidence may provide an idea of relative density.

1. A gannet's (*Sula bassana*) nest with a young bird and an adult. Direct counting of nests is one of the most efficient bird population census methods.

Such findings may also be used to calculate actual densities. Data similar to that described for birds can be gathered for migrating **herbivores** or **carnivores**. All the tracks and feces left by the animals can be counted in a given area during a set period of time.

The **capture and recapture** method is also used to find population density. This method is useful when counting small or fairly immobile animals. It is also useful for counting animals with high population densities in confined habitats.

The capture and recapture method is often used with small mammals, such as rodents, insectivores, reptiles, and amphibians. In this method, individual animals are captured. They are marked and released back to the environment. A second sweep is then made. This sweep turns up several marked animals and other unmarked individuals. The ratio between the first and second groups is considered to be equal to the ratio between the number of individuals captured in the first sweep and the total population.

It is often hard to carry out any kind of census when dealing with living things other than **vertebrates** and trees. In such cases, indirect measurements are made. For example, the amount of **chlorophyll** in a cubic inch of water is useful for calculating the density of a single-celled **algae** population. A grass population can be determined by measuring the amount of oxygen produced through photosynthesis in an area of grassland. These boundaries are proportional to the total number of plants. In turn, this is proportional to the number of individual members, whether they be single-celled algae or blades of grass.

2

2. A diagram of the factors that often control the density of an animal population illustrated through the example of a colony of gannets. The population increases due to the birthrate and immigration, while it diminishes due to emigration and death, shown here by the predation of a great black-backed gull (*Larus marinus*) on a gannet chick. The algebraic sum of the four factors shows, by positive or negative numbers, any increase or decrease in density. In this way, the stability over time is determined for the population in question.

1. A winter survey of birds on Italy's Padana plain. Censuses are used to study the factors that affect the distribution and numbers of organisms in the environment.
2. A census of chamois (*Rupicapra rupicapra*) in a mountain environment.
3. The capture of frogs (*Bufo bufo*) using a plastic sheet during their mating migration is made to carry out a specific census. There are many different census methods and techniques. These range from the direct counting of large animals or tiny eggs suspended in water to the creation of ways to count populations that are difficult to observe.

2

3

4. A direct census of aquatic birds on a southern European marsh. In Europe, continent-wide censuses are organized for some groups of hunted animals, such as ducks and coots. The counts are used to set limits on the total numbers of animals that may be killed.
5. Elk (*Alces alces*) feces in a Finnish forest.
6. Badger (*Meles meles*) feces in a broadleaf forest in northern Italy. The repeated observation of feces is a common way of measuring a population. It is most useful in the case of mammals that are active at night and forest dwellers.

NATURAL RECYCLING

1

2

Nutrient Cycles

The Earth, its atmosphere, and all its ecosystems make up a **closed system.** In a closed system, nothing enters the system from outside. Instead, materials are cycled through the system.

The elements that are absorbed and released by living things enter **nutrient cycles.** These cycles involve both nutrition and decomposition. Nutrition is the process by which living things take in, change, and use the chemicals needed for their survival. Decomposition is the process that puts nutrients, the elements living things need for growth, energy, and repair, back into the cycle. Bacteria play a central role in the nutrient cycle. They are largely responsible for decomposition.

Nitrogen is a needed part of **amino acids** and **proteins.** In the **nitrogen cycle**, proteins are changed into **inorganic nitrogen.** This process occurs in a series of stages and involves the actions of different bacteria. Each type of **decomposer** carries out a certain type of chemical reaction. At the end of the decay process, ammonia is released as an end product. The nitrogen in the ammonia is then changed twice. First, it is changed into a **nitrite.** This is done by bacteria of the *Nitrosomona* genus. Next, it is changed into a **nitrate** by bacteria called *Nitrobacters.*

The most important part of the nitrogen cycle is the making of nitrates. This form of nitrogen can be used by plants. The rate at which nitrates are made directly affects the productivity of the environment.

Nitrogen may leave the cycle through the action of the bacteria that remove nitrates or nitrites. This nitrogen returns to the cycle due to the action of blue-green bacteria. These organisms "fix" the nitrogen in the air. When nitrogen is fixed, it is changed into a form of nitrogen that plants need. In some cases, the bacteria, *Rhizobium*, live in **symbiosis** with higher plants. Symbiosis is an interaction of two species that benefits both species.

Phosphorus also cycles in the environment. Bacteria are involved in the **phosphorus cycle.** They change organic phosphorus into **phosphates.**

In the **sulfur cycle**, bacteria change sulfur into hydrogen sulfide gas. This sulfur gas can rise up from **sediments** deep in the Earth.

The cycles of other elements have different characteristics. For example, metallic elements are not strongly bound in organic compounds. They have **sedimentary cycles.** In sedimentary cycles, elements tend to drop out of the ecosystem through **leaching**, the washing away of substances. They return through the crumbling and dissolving of rocks and through dust, rainfall, and organic debris.

In forest environments, the effects of leaching are lessened by plant cover. Some metals, such as **potassium**, are released into the soil from leaves. These metals may return to the cycle and be reabsorbed. In oak

Pages 22–23
1. A fallen pine tree (*Pinus sylvestris*). Tunnels have been bored into the wood by insects. **2.** Decomposition is already well advanced in a chestnut (*Castanea sativa*). The decomposer organisms — bacteria, blue-green bacteria, and fungi — perform an immense amount of work. An average deciduous forest may form a layer of detritus and dead leaves 40-centimeters (16-inches) deep each autumn.

The nitrogen cycle. About 78 percent of the atmosphere is composed of nitrogen gas. For nitrogen to be useful, it must be "fixed," or changed, by certain organisms. These organisms include blue-green bacteria, microorganisms that live in water or in thin gelatinous strands on dry land. During thunderstorms, flashes of lightning cause a chemical reaction that changes atmospheric nitrogen into the gas nitrogen dioxide. The nitrogen dioxide is washed into the soil by rain. Here we see a *Nostoc* community, blue-green bacteria, that fix nitrogen. **a.** Soil bacteria, called *Nitrobacters*, change the nitrogen from the air into soil nitrates. **b.** Urine, feces, and animal and plant remains are changed by microscopic fungi, called zygomycetes, into ammonia. The ammonia is then changed into nitrates by other bacteria. **c.** *Nitrobacter winogradskyi* is the first phase of the nitrogen cycle, which takes place because of bacteria and organisms active in changing ammonia into nitrates and subsequently nitrites. Some of the nitrates in soil are taken in by the roots of plants. The plants change them into proteins that can be used by animals. The remaining nitrates are dissolved in water in the soil and decomposed by denitrifying bacteria. These bacteria release nitrogen gas that returns to the atmosphere. Plants and higher animals continuously change the inorganic nitrogen of the soil, or nitrates, into complex forms of organic nitrogen, such as proteins. Microscopic organisms return the nitrogen to the soil and atmosphere.

The Nitrogen Cycle

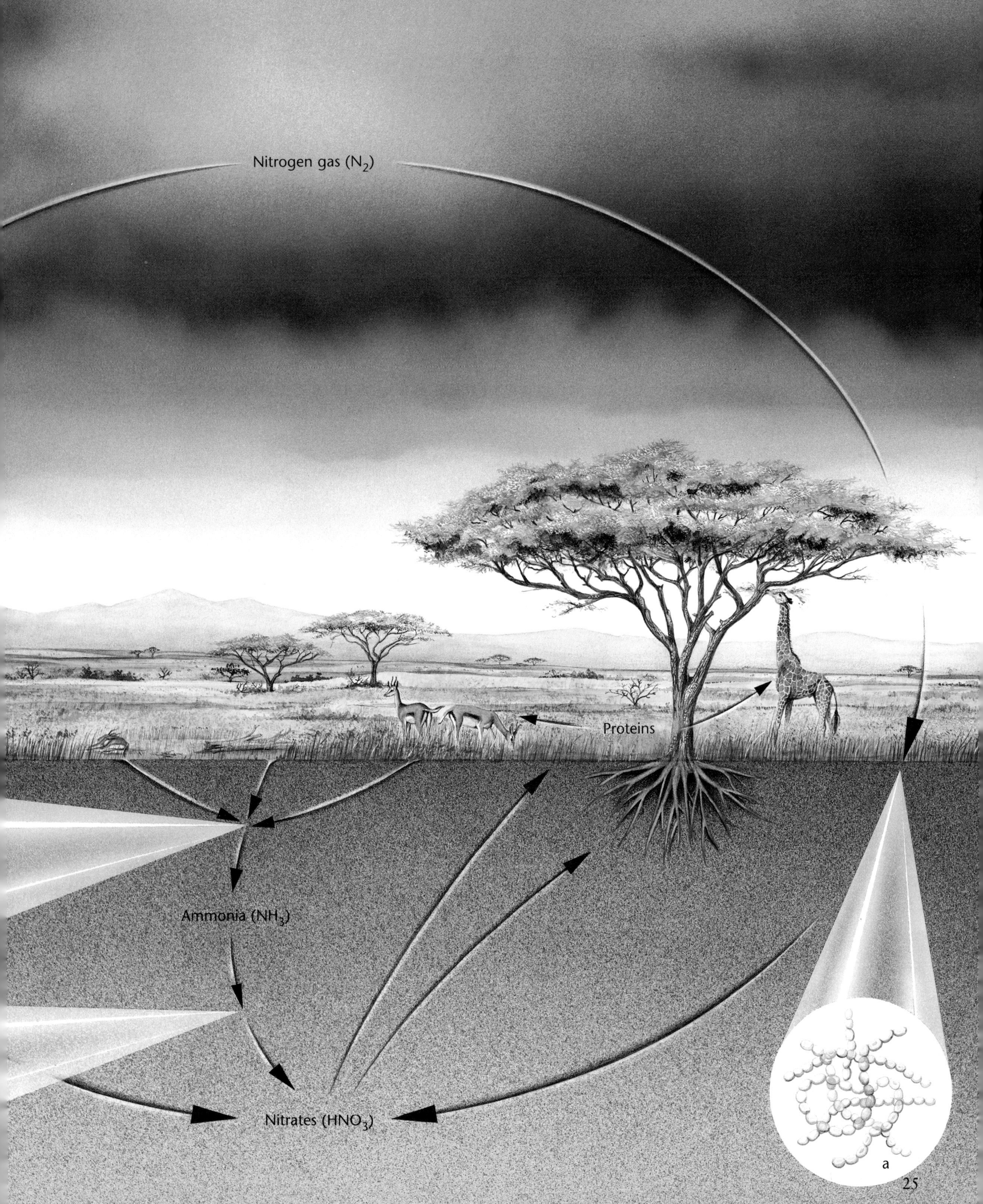

Nitrogen gas (N_2)

Proteins

Ammonia (NH_3)

Nitrates (HNO_3)

a

1

forests, soil potassium comes mainly from the leaves of the trees. In pine forests, it comes from the ferns of the forest undergrowth.

Symbiosis Between Fungi and Plants

Plants get the substances they need from soil. This process is an important part of the nutrient cycle. **Fungi** often play a part in this process. A common partnership, or symbiosis, occurs between some plants and fungi. This relationship is called **mycorrhiza symbiosis.** In this partnership, the underground part of the fungus invades the plant's roots. The fungus helps the plant absorb nitrogen and phosphorous. In exchange, the fungus receives oxygen and water.

Often, a fungus penetrates only the surface of the root. This type of symbiosis is called **ectomycorrhiza symbiosis.** When the root is surrounded by fungus cells, the root lacks the fine hairs that are used to absorb water and mineral salts. This type of symbiosis occurs in almost all the trees of the temperate regions.

A number of well-known components of the detritus chain: **1.** A mold, a microscopic fungus of the class Ascomycetes, on a piece of cheese. **2.** Lichens, associations of fungi and algae, on stones. **3.** Lichens on a larch trunk. **4.** Macroscopic fungi (*Pholiota destruens*) on a felled poplar trunk. **5.** Fungi in an Asian tropical forest. **6.** Parasol mushrooms (*Lepiota procera*) in a forest.

2

3

4

5

6

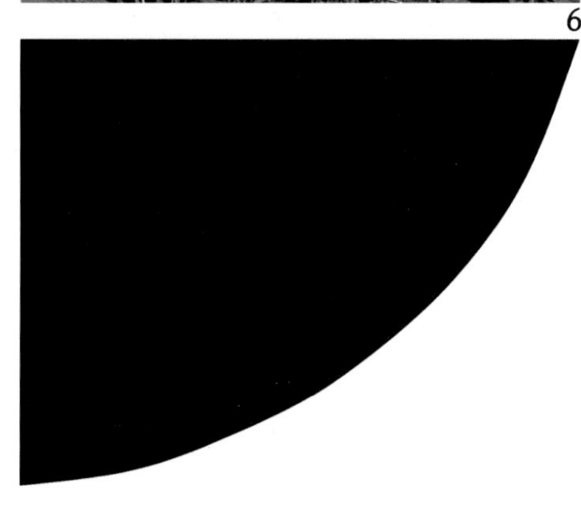

In some cases, the fungus forces its way through the root's surface. The fungus enters the cell structures of the plant. This type of symbiosis is called **endomycorrhiza symbiosis.** Inside the root, the fungus grows in masses. This is the most common type of symbiosis. It occurs in orchids, heathers, grasses, and mosses. Scientists believe that mycorrhiza symbiosis involves more than 90 percent of the families of higher plants.

Young plants grow in proportion to the amount of phosphorus they have in their seeds. Growth stops when there is little or no phosphorus. The partnership between a root and a fungus begins soon after germination. In some plants, if the fungus is not present, growth does not occur.

The partnership between fungi and plants is often specific. So, one fungus species may interact with only one plant species.

Mushrooms are actually the fruiting bodies of a type of fungi. The fungi usually belong to the most complex **Basidiomycete** class. These fungi reproduce by producing **spores.** The spores are released from the **gills** of the cap.

The Energy Balance

In a mature ecosystem, the energy balance between photosynthesis and respiration usually works out evenly. So, the amount of organic matter made in the ecosystem is about equal to the amount used. This occurs mainly through the **detritus** chain. Detritus is the dead organic matter of an ecosystem.

Herbivores and carnivores are consumers. Consumers are organisms that feed on other organisms. However, with respect to the detritus cycle, herbivores and carnivores are also **producers**. These animals directly affect the availability of their food sources by actively seeking them out and eating them. However, the **detritovores** consume organic substances that become available regardless of their actions. Detritovores feed on detritus. Detritus includes the bodies of dead organisms, dead leaves, and the feces and dead hair and skin of animals. There is a lot of detritus available continuously.

Decomposition takes place on a large scale in the soil of a forest. This process involves more than 80 percent of the substances in the whole ecosystem. With decomposition, the nutrient cycle is completed. Organic substances are changed back into their inorganic state. They then enter the nonliving parts of the ecosystem—the air, water, and soil.

Decomposition

Decomposition often starts with the action of microorganisms. These microorganisms may live in the cells of dead tissues. Decomposition may also begin in the digestive systems of the animals that have eaten the dead matter. Once inside the digestive system of an animal, the **microorganisms** take in substances such as sugars and amino acids.

The first organisms to attack soil detritus are bacteria and fungi. The bacteria include nitrifiers such as the *Nitrosomonas*. The fungi include microscopic forms of molds, such as *Penicillium*. Bacteria and fungi that feed on dead matter are often called **opportunistic species**. They have been given this name because they are quick to take advantage of new opportunities. They multiply rapidly. Then they are replaced by other organisms that can digest tough substances, such as **cellulose, lignin, chitin,** and **keratin**.

Decomposition is usually slow. The rate at which the process takes place often depends on how long it takes the fungi to penetrate the plant's **cell walls**. This process involves fungi such as **Ascomycetes, Basidiomycetes,** and **Actinomycetes**. Fungi of these classes grow slowly and have a specialized **metabolism**. When the fungi pierce the plant's cell walls, the fungi create a path that can be used by other microorganisms to penetrate the cell. These microorganisms may include *Clostridia* and *Nitrobacters*. Each species that enters the dead matter may then take part in decomposition.

1

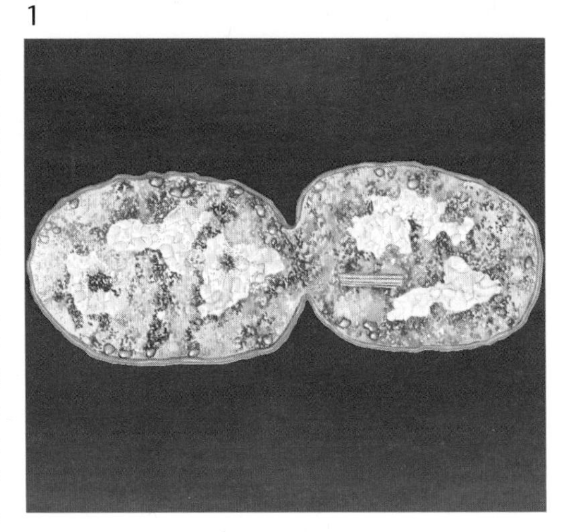

1. A bacteria that fixes atmospheric nitrogen, *Nitrobacter vinelandii*, common in garden soil. This example, which measures about 0.0015 millimeters (0.00006 inches), is in the process of dividing into two cells.

2. A boletus mushroom (*Boletus badius*), a large fungus belonging to the Basidiomycete class, on a forest floor.

2

The Detritus Chain

The work of decomposers is made easier by the **invertebrate** animals that feed on dead organisms. These invertebrates include mites, earthworms, wood lice, millipedes, snails, and slugs. The **larvae** of insects such as **dipterans, coleopterans,** or beetles also break down organic matter.

Invertebrates that feed on dead organic matter break the matter into smaller pieces. When these animals eat detritus, they also eat the **microflora** that live in the detritus. They often depend on these microflora to digest cellulose. Of course, the total number of microorganisms living in forest soil is far greater than the number of living things that eat detritus. However, their mass is only 5 to 10 times greater. The microorganisms are present in greater numbers because they multiply rapidly.

Decomposition can be a very slow process. For example, the complete decomposition of a pine needle on a forest floor may take as long as eight years. The speed of decomposition of leaves depends on the amount of lignin they contain. The amount of lignin is higher in **conifers** than in other types of plants.

The decomposition rate is also linked to temperature. In a tropical forest, decomposition usually occurs quickly. Also, the organic material remains in the ground for a much shorter period of time than it does in the northern coniferous forests.

3. Common broom (*Spartium junceum*) flowering on an Apennine mountainside. Like all members of the legume family, this plant has bacteria in its roots, in this case *Rhizobium.* The bacteria fix atmospheric nitrogen. After taking their place in the root system, these microorganisms produce large quantities of a secretion that forms cysts. The cysts are visible to the unaided eye and are resistant to drying. These are the root nodules, which can be seen in the drawing at the bottom of the page. All of the nitrifying bacteria belong to a single family. Most of them appear to have a flagellum, or whiplike tail, which is used for movement. They are also all organisms that can use oxygen for respiration.

How the Earth Breathes

Plants play a double role in life on Earth. First, they are the most important source of atmospheric oxygen. Second, they also supply, either directly or indirectly, all the organic material on which the survival and growth of all organisms depend. These functions are carried out by both land plants and water plants.

Heterotrophs and Autotrophs

A living organism is a system. The system works by taking in and using molecules that are rich in energy. Many organisms, such as animals, fungi, some **protists**, and most bacteria, are **heterotrophs**. Heterotrophs eat other organisms. In this way, heterotrophs obtain the organic substances they need to survive.

Plants, some protists, and a few species of bacteria differ from the heterotrophs. These organisms can use the energy in sunlight to produce their own food. The food is in the form of sugars that are made from widely available and much simpler molecules, such as **carbon dioxide** and water. Organisms that produce their own food are called **autotrophs**. The term *autotroph* means "self-feeder." So, autotrophs are organisms that make their own food. The process most autotrophs use to make their food is called photosynthesis.

Almost all living things depend on the photosynthesis carried out by autotrophic organisms for their basic supply of energy. Without photosynthesis, the reserves of organic material would soon be exhausted. Life on Earth would cease.

Photosynthesis is a constructive process. During photosynthesis, complex substances are synthesized, or put together, and energy is used. In contrast, **cellular respiration** is a destructive process. In this process an organism eats other living things to get the materials needed to produce energy. In cellular respiration sugars are broken apart, and carbon dioxide and water are released. Cellular respiration occurs in most organisms. Breathing and cellular respiration are not the same process. However, breathing is closely linked to cellular respiration, because breathing provides the oxygen needed for the respiration process.

1. A peat bog where there has recently been a fire. Fire uses oxygen to burn and gives off carbon dioxide. In this way, fire helps recycle elements in ecosystems.
2. The carbon cycle. The burning of dry plant material in a stand of pines in southern Europe (shown left) releases carbon dioxide (CO_2). This becomes part of the reserves of this compound in the atmosphere, about 0.03 percent of all atmospheric gas. The amount of carbon dioxide in the atmosphere also increases as fossil fuels, such as coal and oil, are burned to produce energy, as indicated by the ascending arrow, and by the respiration of organisms. The oil refinery (shown right) processes fossil fuels. Carbon dioxide is removed from the atmosphere and put back into circulation as carbon, in the form of sugars, through photosynthesis. This process is indicated by the descending arrow on the left. The carbon then circulates from plants to animals. It is eventually released again as gas by fungi and bacteria after the death of the organisms.

The Carbon Cycle

Plant Pigments

Photosynthesis takes place as a result of the action of a molecule called chlorophyll. This molecule has a structure similar to that of **hemoglobin**, the red **pigment** in the blood of many organisms that carries oxygen. The structure of chlorophyll differs from hemoglobin mainly because chlorophyll has **magnesium** in place of iron.

In addition to chlorophyll, plants contain other pigments, called **accessory pigments**. They include **carotenoids**, **phycocyanin**, and **phycoerythrin**. The accessory pigments may help to capture the light necessary for photosynthesis. However, they have to transfer the energy they absorb to the chlorophyll so that

3. Autumn in a forest in the Ligurian Apennines. The green color of the chlorophyll is disappearing from the leaves. As autumn advances, other pigments become visible. Late in the season, the yellow-brown colors of tannic acid dominate.

2. Summer. The meadows and forests of the temperate regions are filled with multicolored flowers. The flowers attract pollinating insects. The yellows, reds, and oranges of the flowers are often due to particular pigments called carotenoids.

1. Spring in the Mediterranean region. During spring, trees begin to show signs of life. Pale green leaves appear on the bare branches of the trees.

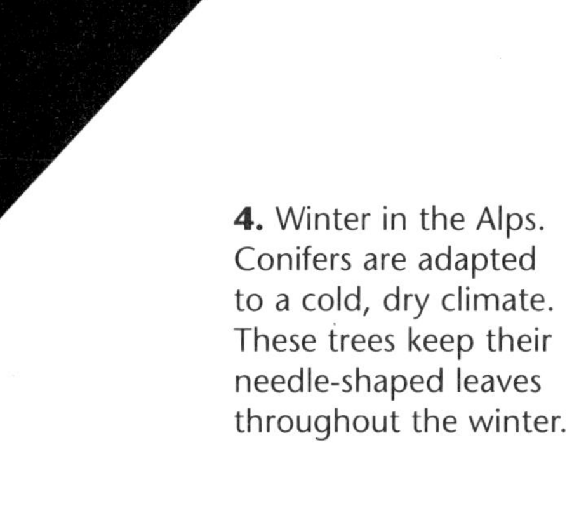

4. Winter in the Alps. Conifers are adapted to a cold, dry climate. These trees keep their needle-shaped leaves throughout the winter.

4

it can be used by the plant during photosynthesis.

The color of a pigment is a direct indication of its ability to absorb light. Chlorophyll absorbs red and blue light very efficiently. But it reflects yellow and green light, which means it uses very little of this light. The accessory pigments may increase the efficiency of the process of photosynthesis because they absorb light of **wavelengths** different than chlorophyll.

Greens, Yellows, and Reds

Deciduous plants are those that lose their leaves each year. In deciduous plants, less photosynthesis takes place as winter approaches. A decrease in photosynthesis is observable by the changing colors of deciduous leaves.

In spring and summer, green is the dominant color of deciduous forests. The chlorophyll in leaves masks the presence of the other pigments. As fall approaches and temperatures drop, production of the green pigment stops. At this point, the yellow pigments, called **xanthophylls**, and the orange pigments, called carotenoids, begin to dominate. The forests take on a whole new color scheme. The pigment **anthocyanin** then becomes dominant. This pigment is responsible for the red tints seen in autumn leaves that have lost all their chlorophyll. The brown color of late autumn and winter leaves is caused by **tannic acid**. This color is a sure sign that the leaves are dead.

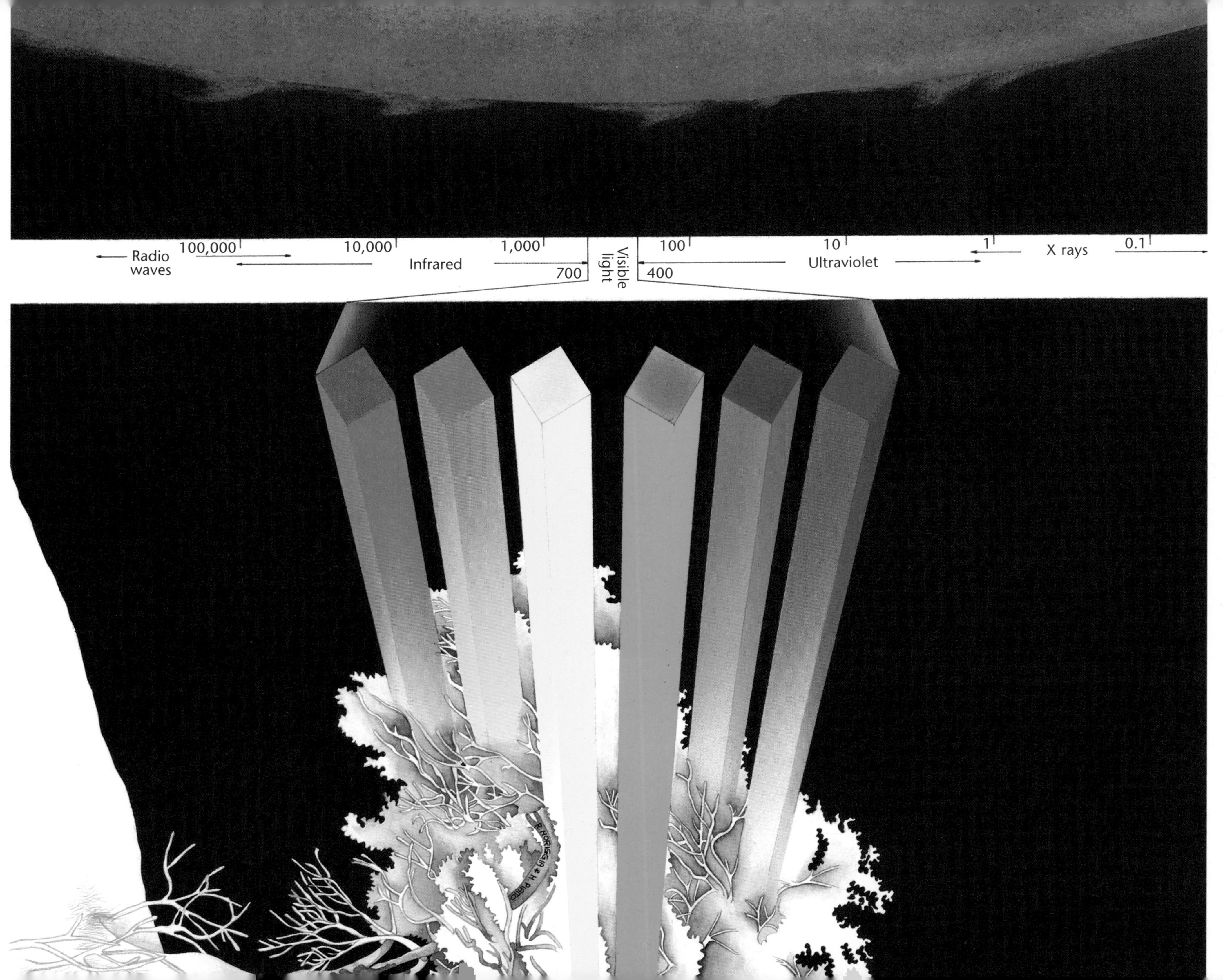

Radio waves
100,000
10,000
Infrared
1,000
700
Visible light
400
100
10
Ultraviolet
1
X rays
0.1

1

1. Visible radiation light has a very restricted wavelength, from 400 to 700 nanometers, compared with total solar radiation, which ranges from little more than 0 to 100,000 nanometers. Ultraviolet radiation has a wavelength less than 400 nanometers. Infrared and radio wavelengths are above 700 nanometers. Plants, like the human eye, are only sensitive to visible radiation. The colors of this radiation range from red to violet. The structure of the pigments contained in leaves is able to absorb most of the visible radiation, except for green light and some yellow light. The green light is all reflected. This action gives plants their characteristic green coloring. On a linear scale, the range of visible radiation would account for just 0.3 percent of all radiation. **2.** A wild hyacinth flower (*Hyacinthus romanus*) **3.** A rose (*Rosa* genus) **4.** Cultivated clematis (*Clematis* genus) **5.** A cultivated pot marigold (*Calendula* genus). The bright colors of flowers come from anthocyanin pigments.

2

3

4

5

The Process of Photosynthesis

Life on Earth depends on the process of photosynthesis. Photosynthesis is a chemical process carried out by plants, some protists, and some bacteria. During photosynthesis, organisms use the energy of sunlight to make food in the form of sugar. The sugar is made by combining carbon dioxide (CO_2) and water (H_2O). Oxygen (O_2), which is needed by most organisms for respiration, is produced as a by-product of photosynthesis.

Because of photosynthesis, nonliving material is constantly "borrowed" from Earth's atmosphere and changed into living material. The living material is formed as organisms that cannot carry out photosynthesis feed on other organisms.

Leaves are the parts of most plants that specialize in photosynthesis. In this way, leaves are natural chemistry laboratories. Many leaves have broad, thin surfaces. This design is an **adaptation** that helps the leaf capture as much sunlight as possible, much like the panels of a solar energy system. The structure of leaf tissues allows gases to be exchanged with the atmosphere. It also allows light to pass through a single leaf to the other leaves below.

Microscopic openings, called **stomata**, are located on the surfaces of leaves. These openings allow the carbon dioxide (CO_2) needed for photosynthesis to be absorbed. They also allow oxygen (O_2) to be released. The oxygen is a by-product of the **oxidation** of water. Leaf stomata can open or close. This movement prevents too much water from being lost through **transpiration** and helps maintain the environment inside the leaf.

These stomata of leaves are arranged in a special way. In plants with leaves that grow vertically, such as the lilies of the forest undergrowth, the stomata are located on both sides of the leaves. In common broadleaf plants, only the underside of the leaves

1. An eighteenth-century engraving of a wild sumac (*Rhus cotinus*). The leaves vary enormously in shape, dimensions, and specific adaptations. However, they all have a function identical to that of solar panels: capturing energy. The energy is then used for the plant's needs.

2. A large leaf of a tree from an Asian tropical forest showing its dense, regular network of veins. The veins allow water to be distributed throughout the leaf's surface.

1

2

3. A section showing the microscopic structure of a leaf. The upper part is protected by the epidermis and a layer of palisade mesophyll cells containing chloroplasts. The lower part is equipped with stomata that allow water (H_2O), carbon dioxide (CO_2), and oxygen (O_2) to be exchanged with the outside environment.

have stomata. There may be thousands of stomata in each square centimeter of leaf.

The upper surface of a leaf is adapted to absorb sunlight. In most leaves, large amounts of chlorophyll exist just below the leaf's upper surface. This is why the tops of most leaves are darker in color than their undersides.

Light and Dark Reactions

Photosynthesis takes place in two phases: the **light reaction** and the **dark reaction**. As its name suggests, the light reaction of photosynthesis requires light and takes place during hours of sunlight. During this phase, visible light is changed into chemical energy. This energy can be used for **cellular metabolism**.

The dark reaction of photosynthesis may occur with or without sunlight. In the dark reaction, the chemical energy produced in the form of **ATP** (adenosine triphosphate) molecules is used to produce sugars.

During the light reaction of photosynthesis, sunlight "excites" an **electron** in the chlorophyll molecule. The electron moves away from its atomic **nucleus**. If the electron moves far enough away, it may be lost. Electrons that are lost or released during the light reaction may enter a long chain of oxidation and **reduction** reactions. These reactions result in the production of high-energy ATP molecules.

The ATP molecules produced in the light reaction tend to break apart easily. However, they can be stored briefly in the plant before being used in chemical reactions that require a supply of energy. During photosynthesis, the most important chemical reactions that occur take place during the dark reaction.

During the dark reaction, the plant makes sugar in the form of **glucose** molecules. These molecules are made using carbon dioxide and water. The ATP produced during the light reaction provides the energy needed to synthesize glucose. The ATP is used for all the plant's other energy needs as well.

ATP also provides the energy for organisms that cannot carry out photosynthesis. These organisms include fungi, animals, **protozoa**, and most types of bacteria. The main difference between the ways that plants and non-photosynthetic organisms produce ATP is the source of energy used to drive the process. Plants produce ATP using solar energy; other organisms use chemical energy.

1. Large burdock leaves (*Arctium lappa*). The broad surface area of some leaves is a visual clue to the process of photosynthesis. However, leaf surface area is actually related to the plant's transpiration needs. The production of chemical energy from sunlight can be performed efficiently by small leaves and evergreen needles.
2. A plant cell. Unlike an animal cell, the plant cell has a cell wall made of cellulose, which strengthens the cell, chloroplasts, and more highly developed vacuoles.

1

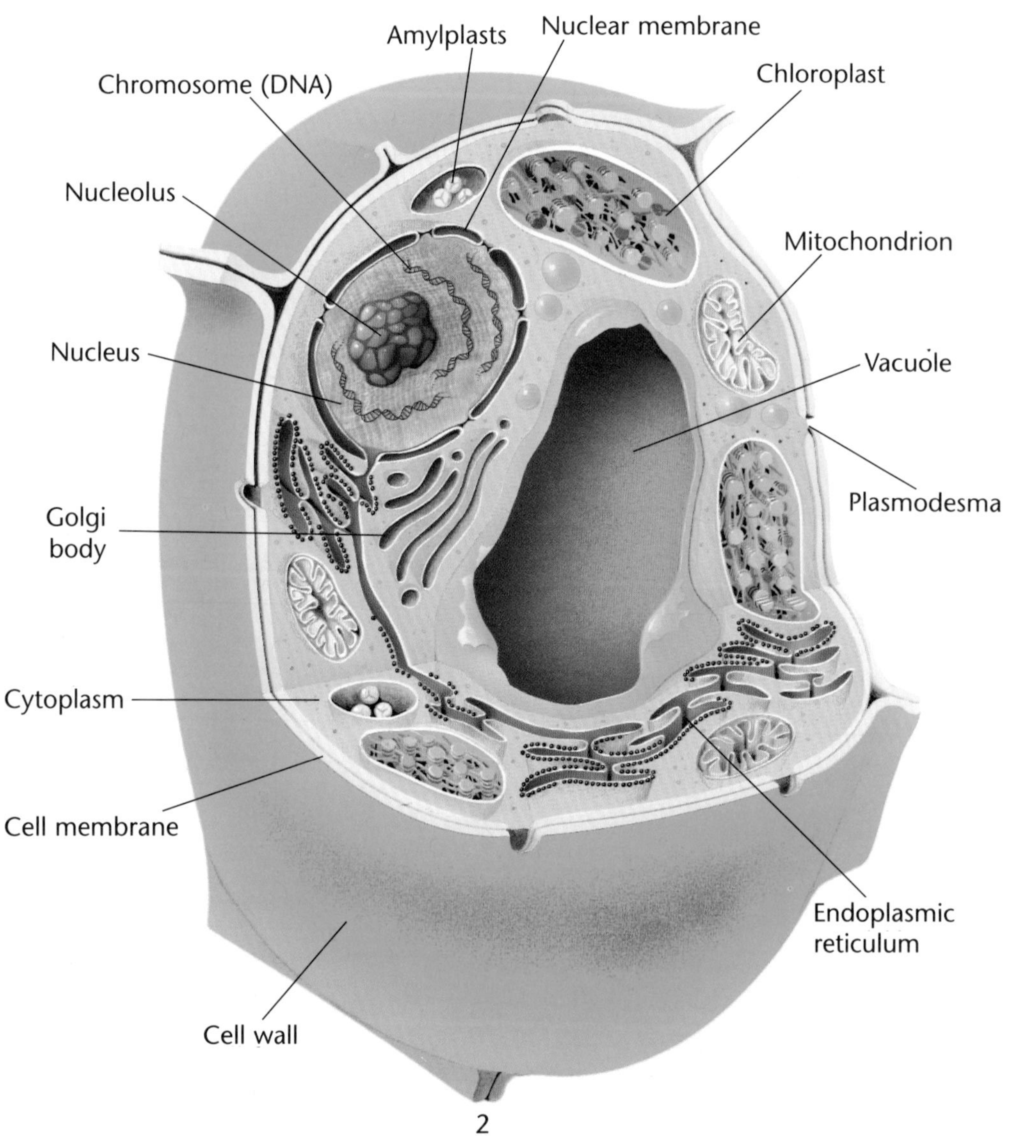

2

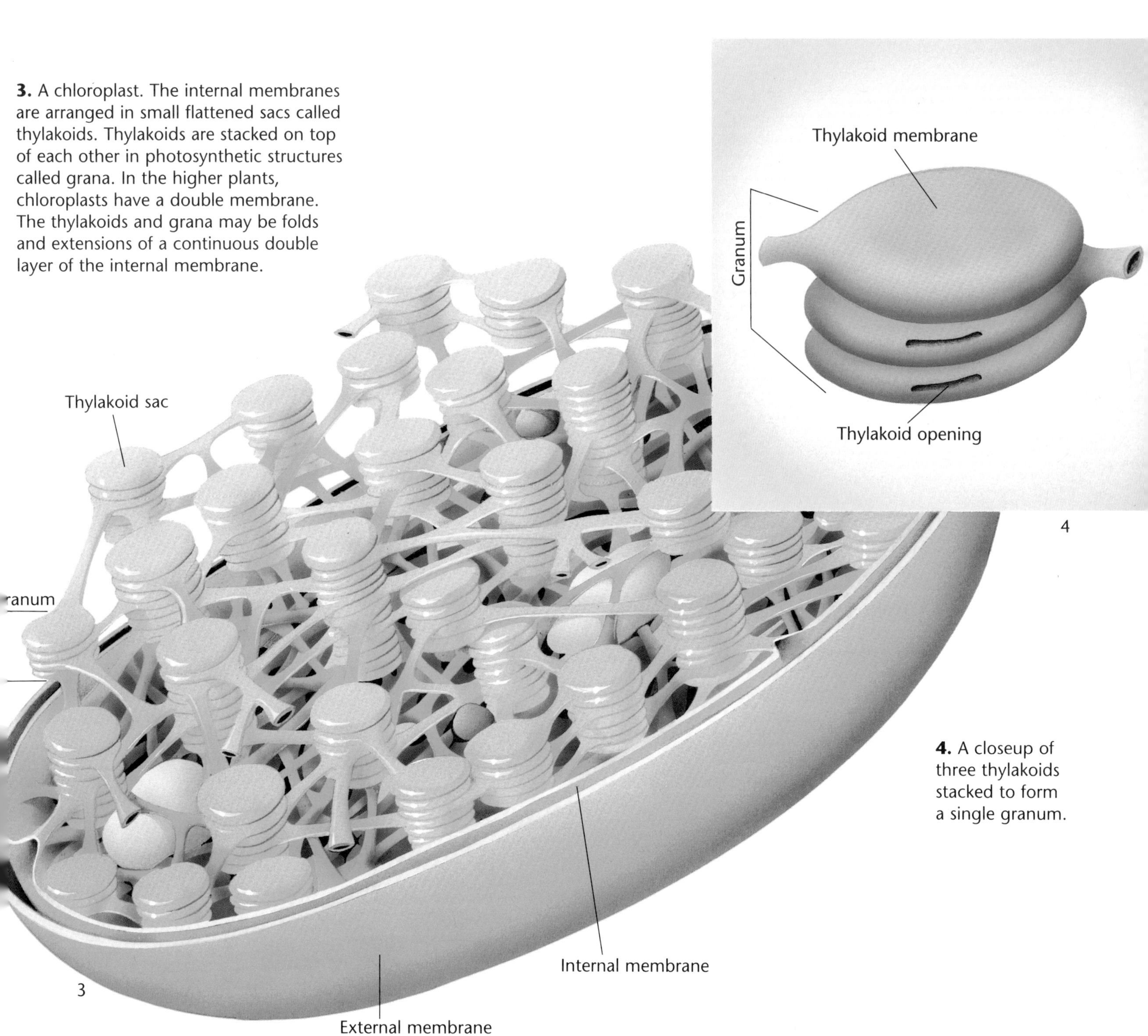

3. A chloroplast. The internal membranes are arranged in small flattened sacs called thylakoids. Thylakoids are stacked on top of each other in photosynthetic structures called grana. In the higher plants, chloroplasts have a double membrane. The thylakoids and grana may be folds and extensions of a continuous double layer of the internal membrane.

4. A closeup of three thylakoids stacked to form a single granum.

The Production of Oxygen

Another essential part of photosynthesis is the production of oxygen. The oxygen is formed by the oxidation of water. Through a long and complex **chain reaction,** water gives off electrons. These electrons are eventually transferred to the chlorophyll, replacing the electrons lost earlier in the process of photosynthesis. In this way the chlorophyll is restored to its original state.

The process of photosynthesis takes place in a number of stages. Together these stages form a cycle. In this cycle chemical compounds are formed, broken down, and then formed again. The carbon atoms obtained from the carbon dioxide in the atmosphere are fixed in the glucose molecules formed during photosynthesis.

Plants use glucose to form more complex **carbohydrates,** such as **sucrose, starch,** and cellulose. These substances act as energy stores that can be kept for long periods and be used and broken down when necessary. A number of intermediate products of the cycle are also used to make the variety of oils contained in most plants. These oils are responsible for the scents given off by leaves, branches, and stems.

Using all the events that occur in photosynthesis, we can now construct a simple equation to summarize the process. On the one hand, the equation illustrates the use of carbon dioxide (CO_2) from the atmosphere and water (H_2O). On the other side of the equation, the **synthesis** of carbohydrates ($C_6H_{12}O_6$) and the release of oxygen (O_2) are shown. These reactions are possible as a result of light falling on an organic structure that has evolved to perform this function to an extraordinary degree of efficiency.

$$6CO_2 + 6H_2O \xrightarrow[\text{Green plants}]{\text{Light}} C_6H_{12}O_6 + 6O_2$$

1

2

4

3

1. Palm leaves capture solar energy and convert it to chemical energy. **2.** Solar panels capture solar energy much like plant leaves. **3.** In contrast, animals like deer obtain chemical energy by consuming organic fuel, such as the carbohydrates in grass. **4.** Machines constructed by humans, like this methane gas-producing station in Venezuela, consume organic fuels for the same reason. **5.** Plants get the energy for their metabolic needs from the sun. This energy is also used to make sugars.

5
Producer
Plant metabolism
Photosynthesis
(solar energy)
Sugars
(high energy molecules)
Primary production
Consumers
Excretion and decomposition

The Chemistry of Life

1

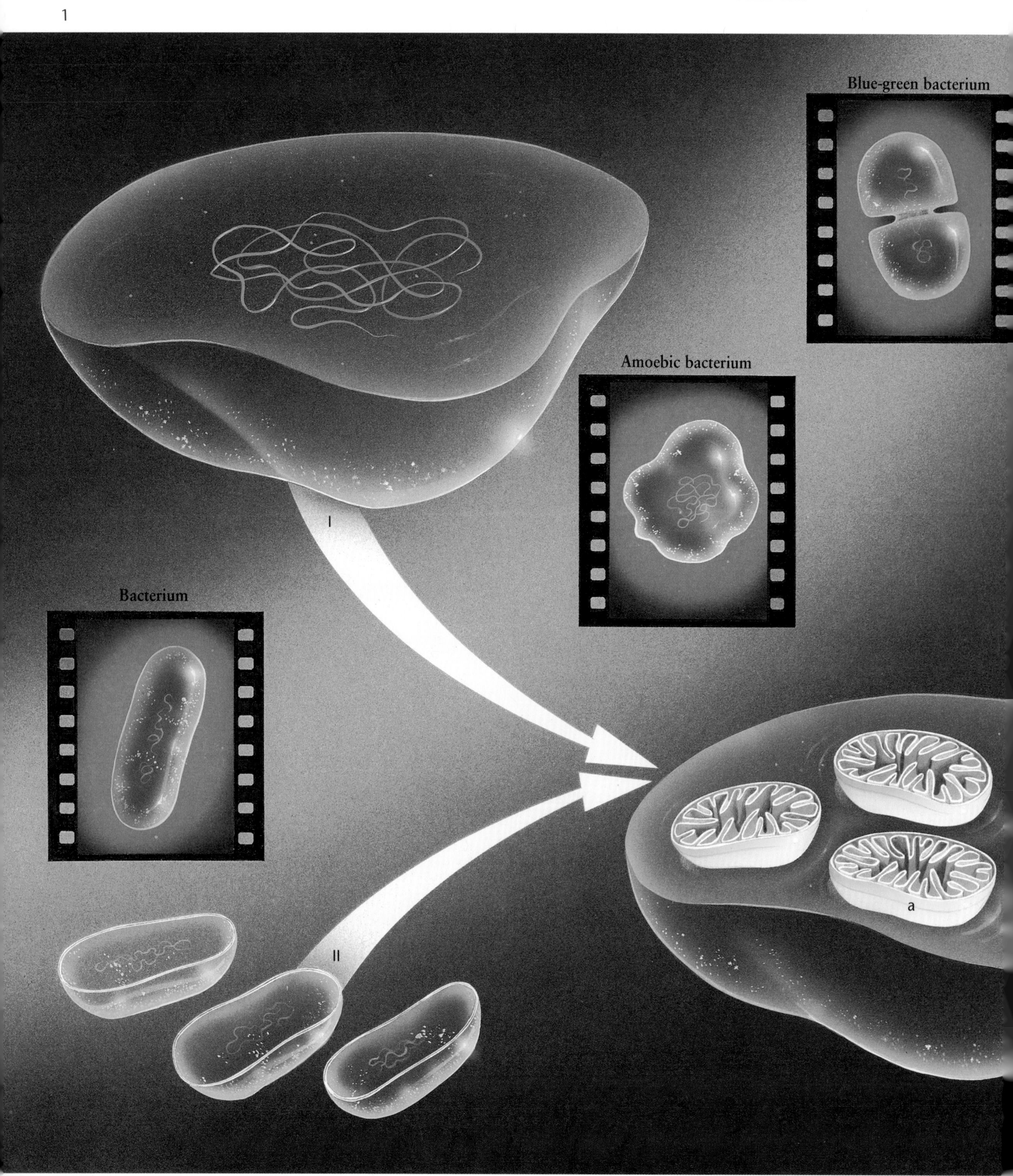

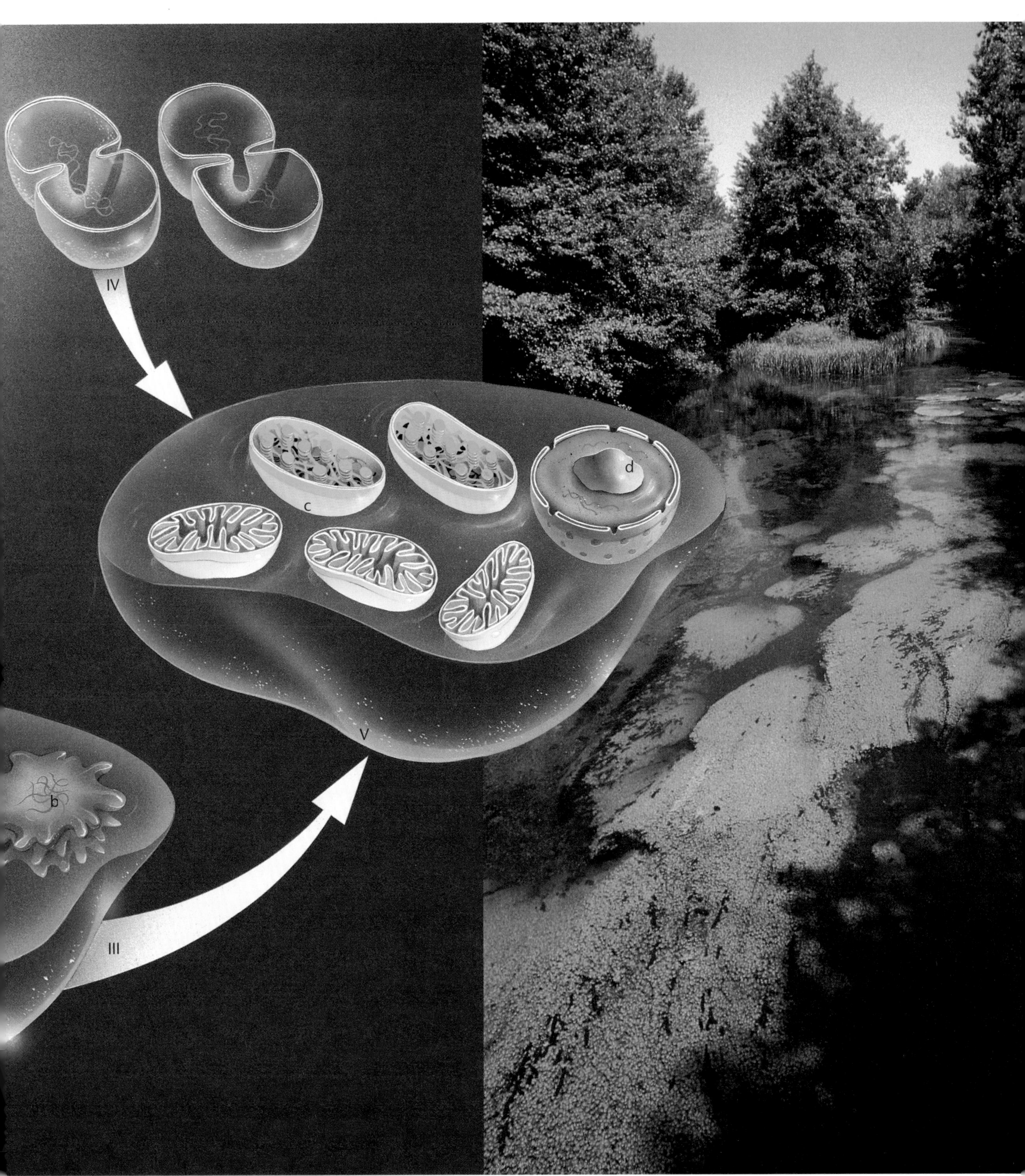

2

1

1. A flock of gannets (*Sula bassana)* in flight on a windy day represents an episode of intense respiration. **2.** The fronds of a palm or the leaves of a water lily (**3**) represent intense photosynthetic activity.

Pages 42–43:
1. Eukaryote plant cells began in a symbiotic relationship. A large amoebic bacterium (**I**) unable to respirate engulfed small bacteria (**II**) capable of respiration. In exchange for raw materials, the small bacteria provided energy and avoided being digested. A primordial cell (**III**) with mitochondria (**a**) and a pre-nucleus (**b**) formed. Later blue-green bacteria capable of photosynthesis (**IV**) united with the primordial cell. This gave rise to the plant cell (**V**) with chloroplasts (**c**), a true nucleus (**d**), and mitochondria.
2. After many millions of years, the primordial cells in an aquatic environment evolved into the plants we know today.

It is hard to imagine what the living world would be like without photosynthesis. This process began in the **primordial oceans**. The primordial oceans were a soup of chemicals and single-celled organisms that could not carry out photosynthesis. These organisms lived far from direct contact with sunlight, at depths greater than 5 meters (16 feet).

In the depths of the oceans, nutrients were scarce. As living things took in larger amounts of food, competition became more intense. By chance, certain bacteria took in molecules that could be stimulated by light energy. These molecules were not digested by the bacteria. Over time, the molecules became part of a system that allowed the bacteria to obtain energy. The energy was available even when nutrients were not.

Oxygen and Aerobic Respiration

As a result of the new system formed by the bacteria, they could now produce oxygen. The oxygen was released by the water molecules the bacteria used to make new organic material. Before this, oxygen had been poisonous to cells. Respiration took place only in the absence of oxygen. Respiration in the absence of oxygen is called **anaerobic respiration**.

When oxygen was first formed by bacterial cells, it was an unwanted by-product. However, the development of systems able to control its strong oxidizing tendency allowed the oxygen to be recycled. This was the second major stage in the evolution of the life-forms we know today. Organisms became capable of **aerobic respiration**. In this process, oxygen became a needed part of a process that used and broke down energy-rich molecules. The energy given off by these molecules was then used by living things in the metabolism and the synthesis of other molecules.

Oxygen began accumulating in the atmosphere. **Ultraviolet rays** changed part of this oxygen into **ozone**. Ozone is a form of oxygen made up of three atoms per molecule instead of the normal two atoms per molecule. The ozone molecules formed a protective layer in the atmosphere. This layer blocked many of the ultraviolet rays harmful to living things. In time, life evolved on dry land as well as at ocean depths.

2

3

Unlike **combustion**, respiration takes place at normal temperatures. The breakdown of organic substances into carbon dioxide and water is a gradual process. This process takes place in a number of stages. Small amounts of energy are released during these stages that are used by the cells.

Respiration involves a complex chain of reactions. The first part of the chain reaction, called **anaerobiosis**, takes place without oxygen. The second part of the chain reaction, which requires oxygen, is called **aerobiosis**. Aerobiosis takes place in small cell **organelles** called **mitochondria**. In the mitochondria, glucose ($C_6H_{12}O_6$) is broken down through oxidation. As with the process of photosynthesis, a general equation illustrates the original nutrient and the end products:

$$C_6H_{12}O_6 + 6O_2 \downarrow 6CO_2 + 6H_2O + \text{ENERGY (36 ATP)}$$

The chemical yield of this complete chain reaction is very high. For each glucose molecule that is broken down, 36 molecules of ATP are formed. About 55 to 60 percent of the energy contained in the glucose molecule can be used by the cells. The remaining energy is lost during the process as waste heat energy.

Prokaryotes and Eukaryotes

Photosynthesis and aerobic respiration are metabolic processes. These processes are closely linked. Both processes are carried out by simple organisms called **prokaryotes**. Prokaryotes are organisms with simple cell structures. Unlike most cells, their cells lack a nucleus and other cell parts surrounded by **membranes**. Bacteria and blue-green bacteria are examples of prokaryotes.

Most multicellular organisms have cells with a distinct nucleus and specialized organelles surrounded by membranes. Organisms with such cells are called **eukaryotes**. The organelles of eukaryote cells carry out a variety of cell functions that keep the cell alive. Plant, animal, fungal, and most protist cells are eukaryote cells.

The development of eukaryote cells from prokaryotes was a great evolutionary step. One of the most interesting and widely accepted theories about the origins of life is that of the symbiotic origins of mitochondria and **chloroplasts**. The mitochondria and chloroplasts are the cell organelles in which eukaryote plant cells carry out respiration and photosynthesis.

According to the theory of symbiotic origins, chloroplasts and mitochondria first developed when organisms similar to **amoebas** engulfed bacteria and blue-green bacteria. A symbiosis, a relationship between two species of organisms that benefits both organisms, began when the prokaryotes were engulfed as food. However, the metabolic functions of these engulfed cells were useful to the organism that had engulfed them. So, instead of digesting the cells, the engulfed cells were kept alive within the **cytoplasm** of the larger cell. A primitive symbiosis began. Later, this symbiosis developed into an inseparable and vital relationship for both cells involved.

Over time, cells became more complex through symbiosis. This allowed living things to take another step forward. They were now self-sufficient. Their cells could now transform, store, and use fair amounts of the energy contained in sunlight.

PRODUCERS AND CONSUMERS

Food Chains and Webs

Living things are linked in various relationships. These relationships are as important to their survival as the physical factors in the environment. Most of these relationships are based on nutrition. From these relationships we can identify **food chains** typical of every ecosystem. Some of these chains are very simple. Others are very complex.

Autotrophs, such as plants, some bacteria, and some protists, change inorganic substances in the environment into organic substances. These substances can be used by heterotrophs, such as animals, fungi, and most protists and bacteria. Plants and other autotrophs are producers. In a forest, there are many food producers—the higher plants in all the forest vegetation layers, mosses, and ferns.

The living things that feed directly on plants are the herbivorous animals. In a food chain, these animals perform the role of **primary consumers**. In the forest, this role is carried out by insects, grain- or seed-eating birds, rodents, and herbivorous mammals, among others.

The primary consumers are a source of food for the **secondary consumers**. These organisms are commonly called **first-order carnivores**. In the forest, this role is carried out by insect-eating birds, reptiles, and carnivorous mammals. In most ecosystems, the last level of the food chain is made up of **tertiary consumers**. Tertiary consumers are also called **second-order carnivores**. These animals feed on first-order carnivores. An example of a tertiary consumer is a bird of prey feeding on a snake.

All the living links in the food chain depend on the previous link. The links also depend on the lowest level of the chain, the producers, and on the process of photosynthesis. If you tried to list all the living things in an ecosystem and trace all the food links between them, using arrows to show the various relationships, you would produce a complex diagram. Your diagram would show all the biological relationships that involve a transfer of energy. This type of diagram is called a **food web**.

A food web diagram is complex. However, it does not give any clue to the number of individual living things at each level. It also does not show the **biomass** involved. Biomass is the total mass of all the living things in an ecosystem.

Food Pyramids

A **food pyramid** is used to show the number of living things at each level in a food web and the total biomass of the ecosystem. Food pyramids show the energy producers at the bottom. The tertiary consumers are shown at the top. Such diagrams are more efficient and easier to understand than food webs.

The number of living things at each level of the pyramid can be used as a unit of measurement. So, the densities of the populations in the ecosystem and the energy relationships between them can be studied. The pyramid can also be used to predict future changes. A pyramid based on biomass provides a lot of useful information.

In a food pyramid, it is easy to see how the great mass of the first level, the primary producers, can support only a small amount, in terms of total mass, of tertiary consumers. Suppose you were to study a fairly simple ecosystem in a restricted area: a willow tree and all the organisms it is capable of supporting. A single willow tree, with a large biomass, can feed many caterpillars, which have a much lower biomass. The caterpillars, in turn, are a source of food for many insect-eating birds. The birds have an even lower biomass than the caterpillars. Lastly, a very small number of tertiary consumers, birds of prey with a very small biomass, can also be supported by this ecosystem.

As food is moved from one level in the pyramid to the next, some energy is lost.

An imaginary ecological pyramid. At the first level are six kinds of primary producers, all plants capable of photosynthesis. All are autotrophs, capable of feeding by changing carbon dioxide and water into the organic compounds needed for their growth. **7.** The leaves and flowers of a willow tree, *Salix caprea* **8.** The branches and flowers of an apple tree, *Malus sylvestris* **9.** The leaves of a maple, *Acer campestre* **10.** A maidenhair fern (*Adiantum capillus-veneris*) **11.** Heather, *Erica carnea* **12.** male fern, *Dryopteris filix-mas.* On the higher levels are heterotrophic organisms that feed on organic substances made by other organisms. The second level of the pyramid contains three kinds of primary consumers, animals that eat plants. **4.** The Eastern chipmunk (*Tamias striatus*) eats mainly nuts. **5.** The grasshopper (*Tettigonia veridissima*) feeds on leaves. **6.** The scarab beetle lives on nectar. On the third level are the secondary consumers, carnivores that feed on herbivores. **2.** The hedgehog (*Erinaceus europaeus*) eats insects and small reptiles. **3.** The Aesculapius' snake (*Elaphe longissima*) feeds mostly on small invertebrates. Lastly, are tertiary consumers that feed on other carnivorous predators. They are shown at the top of the pyramid. **1.** An example is the golden eagle (*Aquila chrysaetos*) a potential predator of the fox.

7

8

9

This energy does not appear on the next level in the pyramid. So, the **energy yield** decreases each time energy is transferred from one organism to another. Some solar energy absorbed by autotrophs — a very small part of the total amount that is available — is used by the autotrophs. The energy that is not used is stored. This energy is passed to other organisms that eat the autotrophs. The higher the organism's position on the pyramid, the fewer the number of individual members that can be supported by an ecosystem.

The impact of different living things on their ecosystems depends on their sizes, population densities, and their ability to change the environment to their own advantage. From this point of view, humans represent an extreme case. Humans are able to erase whole ecosystems and replace them with other artificial systems that are favorable to them. This process began 10,000 years ago. It is the secret behind the success of humans as a species and the birth of extraordinary civilizations. However, if human population growth continues at this rate, it may also come to represent our downfall.

Around ninety percent of the available energy is lost at each level of the ecological pyramid. So, 100 kilograms (220 pounds) of greenery can support no more than 10 kilograms (22 pounds) of plant-eating insects. The insects, in turn, can support no more than 1 kilogram (2 pounds) of hedgehogs and snakes. Eagles can support even less— 100 grams (3.5 ounces) from 1 kilogram (2 pounds) of foxes. So, the birds will be rare not necessarily because they are threatened by people, but because there is not enough food to support them.

1

2

3

4

5

6

10

11

12

1

1. In terms of a food pyramid, the human species is generally located on a level between that of a primary and secondary consumer. A human primary consumer eats cereals, fruit, vegetables, and other plants. **2.** A human secondary consumer eats the meat of herbivorous animals, such as sheep. **3.** However, the position of humans differs from that of most other animals because most people do not live in a natural ecosystem, such as a coniferous forest. Instead, humans gradually try to replace the natural ecosystems, which force the natural ecosystems to maintain low population densities, with completely artificial ecosystems structured to their advantage at the price of additional

3

4

2

energy expenditure. For example, if a coniferous forest is cleared and replaced by fields of barley and potatoes, operations that require energy, as do all the further operations relating to the cultivated crops, the balance of the ecosystem is unfavorable to the wild animals that lived in the forest but favorable to humans, who obtain additional

food resources and can thus increase their population density. **4.** In an ant colony there is no production but only primary or secondary consumption. **5.** The same could be said of human factories, which normally use energy to change products and are certainly incapable of using sunlight to produce food from carbon dioxide.

5

THE DEEP GREEN PLANET

A flying saucer observing the Earth from many points—over the Amazon, Siberia, or Borneo, for example—would see a rolling sea of treetops. Forests are the fundamental environment of our planet. They are the environment that dominates and persists over time, following the colonization of Earth by living things. In a climate that is not too cold or dry, the result of colonization will be a forest of some kind. Whether deciduous broadleaf or evergreen, mixed or coniferous, tropical or temperate, it will still be a forest.

The difference between town and countryside is decided by humans. If nature was left to its own devices, the difference would disappear in a sea of trees, like the famous temples of Cambodia. Apart from the oceans, the tundra, and the deserts, the whole planet is, will be, or would be covered with forest. In a forest environment, the terrestrial eco systems draw breath. Their continuous labors cease, and they finally rest in a stable, durable form that is resistant to change. The concept of the town and the countryside has no future, except in the hands of people who decide it on the basis of their needs. Forests are the true future of the planet. When they are felled, burned, and uprooted, it is that very future that is being destroyed.

Ecologists say that forests are climax environments. Climax environments are stable, well defined, and balanced with a wide range of different species, microclimates, and subsystems. The destruction of forests means destroying something that was established and lasting, and replacing it with something new, unstable, and temporary. The destruction of even small areas of a large forest means the extermination of native species. It also means the wiping out of small worlds with unique characteristics. The destruction of the world's great forests at the current rate—150,000 square kilometers (58,000 square miles) each year—means the destruction of the planet itself. It means increasing the danger of total collapse day by day. Whatever the outcome, our planet without its forests will be a different world: one languishing in the memories of its past splendors. Let's hope that terrible day is still a long way off, and better still, that it never arrives.

RENATO MASSA

Glossary

accessory pigments Color-producing substances in leaves that can capture light but not use it directly

Actinomycetes Single-celled fungi

adaptation A process in which a species changes with environmental conditions

aerobic respiration Respiration using oxygen

aerobiosis The part of the respiration chain reaction that requires oxygen

algae Single-celled and many-celled plantlike organisms with rootlike structures that photosynthesize. They are classified as protists.

amino acids Organic compounds containing nitrogen that link together in long chains to form proteins

amoeba A single-celled protozoan

anaerobic respiration Respiration using molecules other than oxygen; The breaking down of glucose into lactic acid with a modest energy yield (2 ATP)

anaerobiosis The part of the chain reaction of respiration that takes place without oxygen

anthocyanin A blue and red accessory pigment of plants

Ascomycetes Fungi with microscopic or very small fruiting bodies. A saclike structure protects the spores.

association A plant community that covers a wide area made up of a certain population of species with a characteristic appearance and habitat and with a stable duration

ATP Adenosine triphosphate; A chemical formed in the mitochondria of cells that stores energy for use within the cell

autotrophs Organisms that can make their own food

Basidiomycetes Fungi that have a stalk and a cap, such as mushrooms or toadstools

biomass Total weight or volume of a population living in an ecosystem

biotic factors Living parts of an ecosystem

capture and recapture Census method that uses sampling and marking

carbohydrate A chemical containing carbon, hydrogen, and oxygen. Carbohydrates are major nutrients needed by all living things. Sugars and starches are carbohydrates.

carbon dioxide A colorless gas produced by the burning of organic substances and by respiration

carnivores Animals that feed upon other animals

carotenoids Accessory pigments of plants with colors ranging from yellow to orange-red

cell wall Part of plant cells and photosynthetic bacterial cells that surrounds the cell and gives it shape

cellular metabolism All the life processes that take place within the cell and keep the cell alive

cellular respiration Process in which cells use oxygen to break down glucose to obtain energy

cellulose Support substance in many plants that makes up most of the cell walls and gives strength to leaves, roots, and stems. It is the main component of wood.

census A count of the number of individuals of a species in a given area

chain reaction A reaction that triggers a series of other reactions

chitin A rigid material that forms the external skeleton of insects, spiders, and sea creatures with a hard outer covering

chlorophyll Main pigment that colors the green parts of plants. It is needed for the plant to make food by photosynthesis.

chloroplast Tiny structures containing chlorophyll that are part of plant cells and enable the plant to photosynthesize

closed system System in which materials are continuously cycled from within the existing system

coleopteran Insect group that includes beetles and weevils

combustion A chemical process in which a substance reacts rapidly in the presence of oxygen and gives off heat and light; burning

community All the populations living in the same area

conifers Trees, such as pines and firs, that produce uncovered seeds in cones

cytoplasm Cell material located between the cell membrane and the nucleus

dark reaction Portion of the photosynthesis process that can take place with or without light

deciduous forest Environment in which the dominant species are trees that shed their leaves in winter and grow new leaves in spring

decomposers Organisms that break down dead organic matter into simpler substances by chemical or physical means, returning useful substances to the environment

defense mechanism A method used by an organism to protect itself from enemies

detritovores Organisms that feed on organic waste matter

detritus The dead organic matter in an ecosystem

dipterans Insect group including the true flies that have two wings

dominant species The most common species among those present in an area

ecological equivalents Species in different geographical areas that use resources in exactly the same way

ecology The study of the relationships between organisms and their environments

ecosystem All the living and nonliving parts of an environment, which interact to produce a stable system

ectomychorrhiza symbiosis An association of plants and fungi in which threadlike structures wrap around the roots of plants. In this case, the roots lose their hairs and become rounded and large.

edaphic climax Final community in succession that is determined by the physical or chemical makeup of the soil in a particular area

electron Subatomic particle located outside the nucleus that carries a negative charge

endomychorrhiza symbiosis A kind of symbiosis in which threadlike structures enter the tissues and cells of the host plant

energy yield Amount of energy transferred from one organism to another

epiphytes Plants that live on other plants but do not feed on them. They are equipped with air roots or put down roots in materials that gather in cracks on the trunk of the host plant and absorb humidity from the atmosphere.

eukaryotes Organisms made up of cells having a clearly defined nucleus surrounded by a membrane and mitochondria, chloroplasts, and endoplasmic reticulum

feces The solid wastes excreted by organisms

first-order carnivores The first level of consumers that eat meat

food chain A series of organisms through which energy, as food, passes in an ecosystem

food pyramid Diagram showing the number of living things involved in the feeding relationships at each level in the food web and the biomass involved at each level

food web Complex diagram showing all the possible feeding relationships in an ecosystem

formation The largest unit in a community composed of two or more associations

fungi Kingdom of single-celled and many-celled organisms that have nuclei and are unable to carry out photosynthesis. Mushrooms are fungi.

germination The initial stages in the growth of a seed

gill A radiating plate on the bottom of a mushroom cap

glucose A white sugar in crystal form. Glucose is a carbohydrate. It is also the end product of photosynthesis.

grassland Environment with a fairly dry climate in which the dominant plant species are grasses

habitat The particular place an organism lives

hemoglobin A protein containing iron found in the blood of many animals

herbivore An animal that eats only plants

heterotrophs Organisms that obtain their nutrients by eating other organisms

hydrogen sulfide gas A poisonous gas formed when two parts of hydrogen combine with one part of sulfur

inorganic nitrogen Various forms of nitrogen found in the physical environment, such as the elementary gas, nitrous acid (nitrite), nitric acid (nitrate), and ammonia

invertebrate An animal that lacks a backbone

keratin A tough material, made of protein, contained in skin and nails

larva The juvenile stage in the life cycle of most animals without backbones in which an organism hatches from the egg and is unlike the adult form of the animal

leaching A process in which minerals are dissolved or washed out of soil

liana A kind of woody vine that can be found particularly in the tropical rain forest and that roots underground

light reaction The first phase of photosynthesis during which light is absorbed by pigments

lignin A gluelike substance found between the cell walls of woody plant tissue

limiting factor Any environmental factor that restricts the presence or growth of a species

magnesium A soft, silvery, metallic element needed by living things

membrane A thin layer surrounding or separating a biological structure

metabolism All the chemical reactions an organism must carry out to stay alive

microclimate Temperature and humidity characteristics relating to a small part of a larger habitat

microflora Microscopic organisms capable of photosynthesis

microorganism Single-celled and very small, many-celled organisms

mitochondria Tiny structures inside many living cells that change food into energy

mycorrhiza symbiosis A mutually helpful relationship that arises between fungi and the roots of plants. The roots provide shelter, food, and water for the fungus. The fungus provides minerals to the plant.

nitrate Nitrogen compound normally present in soil and easily absorbed by plants

nitrite A form of nitrogen produced as a by-product of ammonia during decomposition

nitrogen cycle The circulation of nitrogen among soil, water, air, and living things

nucleus Large structure present in most cells that controls all cell activities

nutrient Chemical substance living things need for growth, energy, and repair

nutrient cycles Continuous movement of chemical substances from the nonliving environment, through living things, and then back to the environment through the process of decomposition

opportunistic species Organisms, such as bacteria and fungi, that take advantage of new opportunities in an ecosystem

organelles Tiny structures within a cell that help make it work

organic Containing carbon

organism Any living thing

oxidation A chemical reaction involving oxygen

ozone A dark blue gas that is produced in the atmosphere by the action of the sun's rays on oxygen. It absorbs ultraviolet rays from the sun, which can be harmful to organisms.

parasitism Relationship in which one organism lives in or on another and causes harm to it

phosphates Salts formed from the element phosphorus

phosphorus A nonmetallic element needed by living things to make ATP

phosphorus cycle Continuous movement of phosphorus between the living and nonliving factors in an environment

photosynthesis Process by which green plants and some other organisms use the energy in sunlight to combine carbon dioxide and water to produce food in the form of glucose

phycocyanin Blue accessory pigment of plants

phycoerythrin Red accessory pigment of plants

pigment A substance that colors the tissues or cells of animals and plants

population density The number of organisms of a species per unit area

potassium A metallic element needed for the synthesis of proteins in living things

predator An animal that lives by killing and eating other animals

primary consumers Heterotrophic animals known as herbivores that feed directly on the organic substances produced by plants

primordial ocean Earth's early ocean from which the predecessors of today's life forms emerged

producers Autotrophic organisms that produce organic substances from sunlight through photosynthesis

prokaryote Single-celled organisms, such as bacteria, that have cells lacking compartments enclosed by membranes

proteins Large molecules made up of chains of amino acids

protists Mostly single-celled organisms that have nuclei and compartments enclosed by membranes

protozoa Single-celled animallike organisms classified in the kingdom Protista

rain forests Forests in the equatorial and tropical regions, rich in plant species and characterized by hot, wet climates

range of tolerance Total amount of variation within which a species can survive for a given environmental factor

reduction A chemical reaction in which an electron is added to an atom or ion. Reduction occurs when oxygen is removed from a molecule.

regeneration The ability of an organism to regrow lost parts

relative density Measurement of how a population has changed over time or in different environments

sampling Measuring parts representative of certain characteristics of a larger unit

savanna A tropical grassland that ranges in moisture from dry scrubland to wet, open woodland

scrub Vegetation consisting mainly of stunted trees or shrubs

secondary consumers Heterotrophic animals known as carnivores that feed on primary consumers

second-order carnivores Carnivores that feed on other carnivores. They are the second level of carnivores in a food chain.

sediment Sand or stone deposited by water, wind, or a glacier

sedimentary cycles Continuous processes in which soil and rock join to form rocks and then break apart again to form sediments

spore A reproductive cell that can develop into a new organism without first having to join with another cell

starch A carbohydrate in plants

stomata Microscopic openings in leaves through which gases are exchanged with the environment

succession Changes in ecosystems from less stable to more stable forms

succulent A plant that has large stems or leaves for storing water

sucrose Form of sugar formed in beets and sugarcane that is refined to make table sugar

sulfur Yellow, nonmetallic element that makes up the amino acids of living things

sulfur cycle The continuous movement of sulfur and sulfur compounds between the living and nonliving parts of an ecosystem

symbiosis Relationship between two species that live close together with each other, providing benefits (food or protection) to both organisms

synthesis Reaction in which a substance is produced from two or more substances

taiga Wooded environment with longer mild periods than the tundra. The borders of the extensive taiga conifer or birch forests are often sharply defined by the open tundra environment.

tannic acid Substance causing the brown coloring characteristic of dead or dying leaves

territorial climax Final stage of ecological succession

tertiary consumers Heterotrophic animals that feed on other carnivores, the secondary consumers. Tertiary consumers are usually at the top of the ecological food pyramid.

transpiration Loss of water through the external surface of living tissue

tundra Cold arctic environment characterized by frozen earth just a few inches from the surface and a brief, mild summer period. The vegetation is composed mainly of mosses, lichens, and dwarf forms of birch, alder, willow, and conifers.

ultraviolet rays Electromagnetic energy having wavelengths longer than purple light and shorter than X rays

undergrowth Portion of plant cover in a forest that grows low to the ground

vertebrate An animal that has a backbone

wavelengths The distance between two corresponding points on two consecutive waves

xanthophylls Yellow accessory pigments of plants

FURTHER READING

AIT Staff. *Earth, the Environment, and Beyond from Science Source.* Grewar, Mindy, ed. Agency for Instructional Technology, 1992
Baines, John. *Exploring Humans and the Environment.* Raintree Steck-Vaughn, 1992
Better Earth Series. Enslow, 1993
Blashfield, Jean F. and Black, Wallace B. *Global Warming.* Childrens, 1991
Book of the Earth. EDC, 1993
Brimner, Larry D. *Unusual Friendships: Symbiosis in the Animal World.* Watts, 1993
Bronze, Lewis, et al. *The Blue Peter Green Book.* Parkwest, 1992
Brumley, Karen. *Saving Our Planet.* American Education Publishing, 1991
Challand, Helen J. *Disappearing Wetlands.* Childrens, 1992
Ecology and Plant Life. Prentice Hall General Reference and Travel, 1993
Gallant, Roy A. *Earth's Vanishing Forests.* Macmillan Child Group, 1992
Ganeri, Anita. *Ponds and Pond Life.* Watts, 1993
———. *Ponds, Rivers, and Lakes.* Macmillan Child Group, 1992
Hester, Nigel. *The Living River.* Watts, 1991
Leinwand, Gerald. *The Environment.* Facts on File, 1990
Peacock, Graham and Hudson, Terry. *Exploring Habitats.* Raintree Steck-Vaughn, 1992
Raintree Steck-Vaughn Staff. *Atlas of the Environment.* Coote, Roger, ed. Raintree Steck-Vaughn, 1992
Tesar, Jenny. *Endangered Habitats,* Facts on File, 1991
Walker, Jane. *Vanishing Habitats and Species,* Watts, 1993

INDEX

Picture credits

Photographs

The first number refers to the page. The number in parentheses refers to the illustration. U. BÄR VERLAG, Zürich (MAXIMILIEN BRUGGMANN): 50. ADRIANA CASDIA, Milan: 47 (5). DAVIDE CERIOLI, Nicorvo, Pavia: 32 (2), 35 (2). DUILIO CITI, Chiavari: 22, 23, 31 (right), 32 (3), 46 (8), 47 (12), 48 (2), 49 (5), 51 (left). ALBERTO CONTRI, Milan: 10 (second and fifth from top), 26 (2, 3), 32 (4), 35 (5). GIANNI COSTANTINO-PUBLICITUR: 2–3. EDITORIAL JACA BOOK, Milan (MONICA CARABELLA): 21 (4), 27 (4); (LORENZO FORNASARI): 20 (2), 21 (6); (RENATO MASSA): 10 (first, third, fourth, and sixth from top), 12 (2), 16–17 (all), 18 (1), 19 (2), 20 (1), 21 (5), 26 (1), 27 (5,6), 28 (2), 29 (3), 30 (1), 31 (left), 32 (1), 35 (4), 36 (2), 38 (1), 40 (1, 3), 43 (2), 44 (1), 45 (3), 46 (7, 9), 47 (1, 3, 4, 6, 10, 11), 48 (1, 3), 49 (4); (CARLO SCOTTI): 8; (MIREILLE VAUTIER): 51 (right). ENI, (Italian National Hydrocarbons Council), Rome: 40 (4). GIOVANNI GIOVINE, Bergamo: 20 (3). WALTER GRANERI, Milan: 13 (3), 47 (2). GIOVANNI PINNA, Milan: 45 (2). FABIO TERRANEO, Giussano, Milan: 40 (2).

Color plates and drawings

EDITORIAL JACA BOOK, Milan (REMO BERSELLI): 19, 37; (MARIA ELENA GONANO): 10–11, 14–15, 24–25, 38–39, 42–43; (ROSALBA MORIGGIA & MARIA PIATTO): 29, 34–35, 41; (GIULIA RE): 12 (1), 17, 28 (1).

Separate illustrations

pages 2–3: Cuba's Viñales valley at dawn.
page 8: A fragment of rock containing a fossilized *Ginkgo andiantoides* leaf from the lower Paleocene epoch of about 60 million years ago, United States.
page 50: A coniferous forest on Baranof Island in the Alexander Archipelago, Alaska, silhouetted against the late afternoon sun
page 51 left: Autumn in a deciduous broadleaf forest in the Ligurian Apennines, Italy. Right: A tropical forest halfway up the Marañón River, a tributary of the Amazon River